Number Theory: A Very Short Introduction

THE ANIMAL KINGDOM Peter Holland
ANIMAL RIGHTS David DeGrazia
THE ANTARCTIC Klaus Dodds
ANTHROPOCENE Erle C. Ellis
ANTISEMITISM Steven Beller
ANXIETY Daniel Freeman and Jason Freeman
THE APOCRYPHAL GOSPELS Paul Foster
APPLIED MATHEMATICS Alain Goriely
ARCHAEOLOGY Paul Bahn
ARCHITECTURE Andrew Ballantyne
ARISTOCRACY William Doyle
ARISTOTLE Jonathan Barnes
ART HISTORY Dana Arnold
ART THEORY Cynthia Freeland
ARTIFICIAL INTELLIGENCE Margaret A. Boden
ASIAN AMERICAN HISTORY Madeline Y. Hsu
ASTROBIOLOGY David C. Catling
ASTROPHYSICS James Binney
ATHEISM Julian Baggini
THE ATMOSPHERE Paul I. Palmer
AUGUSTINE Henry Chadwick
AUSTRALIA Kenneth Morgan
AUTISM Uta Frith
AUTOBIOGRAPHY Laura Marcus
THE AVANT GARDE David Cottington
THE AZTECS Davíd Carrasco
BABYLONIA Trevor Bryce
BACTERIA Sebastian G. B. Amyes
BANKING John Goddard and John O. S. Wilson
BARTHES Jonathan Culler
THE BEATS David Sterritt
BEAUTY Roger Scruton
BEHAVIOURAL ECONOMICS Michelle Baddeley
BESTSELLERS John Sutherland
THE BIBLE John Riches
BIBLICAL ARCHAEOLOGY Eric H. Cline
BIG DATA Dawn E. Holmes
BIOGRAPHY Hermione Lee
BIOMETRICS Michael Fairhurst
BLACK HOLES Katherine Blundell
BLOOD Chris Cooper
THE BLUES Elijah Wald
THE BODY Chris Shilling
THE BOOK OF COMMON PRAYER Brian Cummings
THE BOOK OF MORMON Terryl Givens
BORDERS Alexander C. Diener and Joshua Hagen
THE BRAIN Michael O'Shea
BRANDING Robert Jones
THE BRICS Andrew F. Cooper
THE BRITISH CONSTITUTION Martin Loughlin
THE BRITISH EMPIRE Ashley Jackson
BRITISH POLITICS Anthony Wright
BUDDHA Michael Carrithers
BUDDHISM Damien Keown
BUDDHIST ETHICS Damien Keown
BYZANTIUM Peter Sarris
C. S. LEWIS James Como
CALVINISM Jon Balserak
CANCER Nicholas James
CAPITALISM James Fulcher
CATHOLICISM Gerald O'Collins
CAUSATION Stephen Mumford and Rani Lill Anjum
THE CELL Terence Allen and Graham Cowling
THE CELTS Barry Cunliffe
CHAOS Leonard Smith
CHARLES DICKENS Jenny Hartley
CHEMISTRY Peter Atkins
CHILD PSYCHOLOGY Usha Goswami
CHILDREN'S LITERATURE Kimberley Reynolds
CHINESE LITERATURE Sabina Knight
CHOICE THEORY Michael Allingham
CHRISTIAN ART Beth Williamson
CHRISTIAN ETHICS D. Stephen Long
CHRISTIANITY Linda Woodhead
CIRCADIAN RHYTHMS Russell Foster and Leon Kreitzman
CITIZENSHIP Richard Bellamy
CIVIL ENGINEERING David Muir Wood
CLASSICAL LITERATURE William Allan
CLASSICAL MYTHOLOGY Helen Morales
CLASSICS Mary Beard and John Henderson
CLAUSEWITZ Michael Howard
CLIMATE Mark Maslin

CLIMATE CHANGE Mark Maslin
CLINICAL PSYCHOLOGY Susan Llewelyn and Katie Aafjes-van Doorn
COGNITIVE NEUROSCIENCE Richard Passingham
THE COLD WAR Robert McMahon
COLONIAL AMERICA Alan Taylor
COLONIAL LATIN AMERICAN LITERATURE Rolena Adorno
COMBINATORICS Robin Wilson
COMEDY Matthew Bevis
COMMUNISM Leslie Holmes
COMPARATIVE LITERATURE Ben Hutchinson
COMPLEXITY John H. Holland
THE COMPUTER Darrel Ince
COMPUTER SCIENCE Subrata Dasgupta
CONCENTRATION CAMPS Dan Stone
CONFUCIANISM Daniel K. Gardner
THE CONQUISTADORS Matthew Restall and Felipe Fernández-Armesto
CONSCIENCE Paul Strohm
CONSCIOUSNESS Susan Blackmore
CONTEMPORARY ART Julian Stallabrass
CONTEMPORARY FICTION Robert Eaglestone
CONTINENTAL PHILOSOPHY Simon Critchley
COPERNICUS Owen Gingerich
CORAL REEFS Charles Sheppard
CORPORATE SOCIAL RESPONSIBILITY Jeremy Moon
CORRUPTION Leslie Holmes
COSMOLOGY Peter Coles
COUNTRY MUSIC Richard Carlin
CRIME FICTION Richard Bradford
CRIMINAL JUSTICE Julian V. Roberts
CRIMINOLOGY Tim Newburn
CRITICAL THEORY Stephen Eric Bronner
THE CRUSADES Christopher Tyerman
CRYPTOGRAPHY Fred Piper and Sean Murphy
CRYSTALLOGRAPHY A. M. Glazer
THE CULTURAL REVOLUTION Richard Curt Kraus
DADA AND SURREALISM David Hopkins
DANTE Peter Hainsworth and David Robey
DARWIN Jonathan Howard
THE DEAD SEA SCROLLS Timothy H. Lim
DECADENCE David Weir
DECOLONIZATION Dane Kennedy
DEMOCRACY Bernard Crick
DEMOGRAPHY Sarah Harper
DEPRESSION Jan Scott and Mary Jane Tacchi
DERRIDA Simon Glendinning
DESCARTES Tom Sorell
DESERTS Nick Middleton
DESIGN John Heskett
DEVELOPMENT Ian Goldin
DEVELOPMENTAL BIOLOGY Lewis Wolpert
THE DEVIL Darren Oldridge
DIASPORA Kevin Kenny
DICTIONARIES Lynda Mugglestone
DINOSAURS David Norman
DIPLOMACY Joseph M. Siracusa
DOCUMENTARY FILM Patricia Aufderheide
DREAMING J. Allan Hobson
DRUGS Les Iversen
DRUIDS Barry Cunliffe
DYNASTY Jeroen Duindam
DYSLEXIA Margaret J. Snowling
EARLY MUSIC Thomas Forrest Kelly
THE EARTH Martin Redfern
EARTH SYSTEM SCIENCE Tim Lenton
ECONOMICS Partha Dasgupta
EDUCATION Gary Thomas
EGYPTIAN MYTH Geraldine Pinch
EIGHTEENTH-CENTURY BRITAIN Paul Langford
THE ELEMENTS Philip Ball
EMOTION Dylan Evans
EMPIRE Stephen Howe
ENERGY SYSTEMS Nick Jenkins
ENGELS Terrell Carver
ENGINEERING David Blockley
THE ENGLISH LANGUAGE Simon Horobin
ENGLISH LITERATURE Jonathan Bate

THE ENLIGHTENMENT John Robertson
ENTREPRENEURSHIP Paul Westhead and Mike Wright
ENVIRONMENTAL ECONOMICS Stephen Smith
ENVIRONMENTAL ETHICS Robin Attfield
ENVIRONMENTAL LAW Elizabeth Fisher
ENVIRONMENTAL POLITICS Andrew Dobson
EPICUREANISM Catherine Wilson
EPIDEMIOLOGY Rodolfo Saracci
ETHICS Simon Blackburn
ETHNOMUSICOLOGY Timothy Rice
THE ETRUSCANS Christopher Smith
EUGENICS Philippa Levine
THE EUROPEAN UNION Simon Usherwood and John Pinder
EUROPEAN UNION LAW Anthony Arnull
EVOLUTION Brian and Deborah Charlesworth
EXISTENTIALISM Thomas Flynn
EXPLORATION Stewart A. Weaver
EXTINCTION Paul B. Wignall
THE EYE Michael Land
FAIRY TALE Marina Warner
FAMILY LAW Jonathan Herring
FASCISM Kevin Passmore
FASHION Rebecca Arnold
FEDERALISM Mark J. Rozell and Clyde Wilcox
FEMINISM Margaret Walters
FILM Michael Wood
FILM MUSIC Kathryn Kalinak
FILM NOIR James Naremore
THE FIRST WORLD WAR Michael Howard
FOLK MUSIC Mark Slobin
FOOD John Krebs
FORENSIC PSYCHOLOGY David Canter
FORENSIC SCIENCE Jim Fraser
FORESTS Jaboury Ghazoul
FOSSILS Keith Thomson
FOUCAULT Gary Gutting
THE FOUNDING FATHERS R. B. Bernstein
FRACTALS Kenneth Falconer
FREE SPEECH Nigel Warburton
FREE WILL Thomas Pink
FREEMASONRY Andreas Önnerfors
FRENCH LITERATURE John D. Lyons
THE FRENCH REVOLUTION William Doyle
FREUD Anthony Storr
FUNDAMENTALISM Malise Ruthven
FUNGI Nicholas P. Money
THE FUTURE Jennifer M. Gidley
GALAXIES John Gribbin
GALILEO Stillman Drake
GAME THEORY Ken Binmore
GANDHI Bhikhu Parekh
GARDEN HISTORY Gordon Campbell
GENES Jonathan Slack
GENIUS Andrew Robinson
GENOMICS John Archibald
GEOFFREY CHAUCER David Wallace
GEOGRAPHY John Matthews and David Herbert
GEOLOGY Jan Zalasiewicz
GEOPHYSICS William Lowrie
GEOPOLITICS Klaus Dodds
GERMAN LITERATURE Nicholas Boyle
GERMAN PHILOSOPHY Andrew Bowie
GLACIATION David J. A. Evans
GLOBAL CATASTROPHES Bill McGuire
GLOBAL ECONOMIC HISTORY Robert C. Allen
GLOBALIZATION Manfred Steger
GOD John Bowker
GOETHE Ritchie Robertson
THE GOTHIC Nick Groom
GOVERNANCE Mark Bevir
GRAVITY Timothy Clifton
THE GREAT DEPRESSION AND THE NEW DEAL Eric Rauchway
HABERMAS James Gordon Finlayson
THE HABSBURG EMPIRE Martyn Rady
HAPPINESS Daniel M. Haybron
THE HARLEM RENAISSANCE Cheryl A. Wall
THE HEBREW BIBLE AS LITERATURE Tod Linafelt
HEGEL Peter Singer
HEIDEGGER Michael Inwood

THE HELLENISTIC AGE
Peter Thonemann
HEREDITY John Waller
HERMENEUTICS Jens Zimmermann
HERODOTUS Jennifer T. Roberts
HIEROGLYPHS Penelope Wilson
HINDUISM Kim Knott
HISTORY John H. Arnold
THE HISTORY OF ASTRONOMY
Michael Hoskin
THE HISTORY OF CHEMISTRY
William H. Brock
THE HISTORY OF CHILDHOOD
James Marten
THE HISTORY OF CINEMA
Geoffrey Nowell-Smith
THE HISTORY OF LIFE
Michael Benton
THE HISTORY OF MATHEMATICS
Jacqueline Stedall
THE HISTORY OF MEDICINE
William Bynum
THE HISTORY OF PHYSICS
J. L. Heilbron
THE HISTORY OF TIME
Leofranc Holford-Strevens
HIV AND AIDS Alan Whiteside
HOBBES Richard Tuck
HOLLYWOOD Peter Decherney
THE HOLY ROMAN EMPIRE
Joachim Whaley
HOME Michael Allen Fox
HOMER Barbara Graziosi
HORMONES Martin Luck
HUMAN ANATOMY
Leslie Klenerman
HUMAN EVOLUTION Bernard Wood
HUMAN RIGHTS Andrew Clapham
HUMANISM Stephen Law
HUME A. J. Ayer
HUMOUR Noël Carroll
THE ICE AGE Jamie Woodward
IDENTITY Florian Coulmas
IDEOLOGY Michael Freeden
THE IMMUNE SYSTEM
Paul Klenerman
INDIAN CINEMA
Ashish Rajadhyaksha
INDIAN PHILOSOPHY Sue Hamilton
THE INDUSTRIAL REVOLUTION
Robert C. Allen
INFECTIOUS DISEASE
Marta L. Wayne
and Benjamin M. Bolker
INFINITY Ian Stewart
INFORMATION Luciano Floridi
INNOVATION Mark Dodgson and
David Gann
INTELLECTUAL PROPERTY
Siva Vaidhyanathan
INTELLIGENCE Ian J. Deary
INTERNATIONAL LAW
Vaughan Lowe
INTERNATIONAL MIGRATION
Khalid Koser
INTERNATIONAL RELATIONS
Paul Wilkinson
INTERNATIONAL SECURITY
Christopher S. Browning
IRAN Ali M. Ansari
ISLAM Malise Ruthven
ISLAMIC HISTORY Adam Silverstein
ISOTOPES Rob Ellam
ITALIAN LITERATURE
Peter Hainsworth and David Robey
JESUS Richard Bauckham
JEWISH HISTORY David N. Myers
JOURNALISM Ian Hargreaves
JUDAISM Norman Solomon
JUNG Anthony Stevens
KABBALAH Joseph Dan
KAFKA Ritchie Robertson
KANT Roger Scruton
KEYNES Robert Skidelsky
KIERKEGAARD Patrick Gardiner
KNOWLEDGE Jennifer Nagel
THE KORAN Michael Cook
KOREA Michael J. Seth
LAKES Warwick F. Vincent
LANDSCAPE ARCHITECTURE
Ian H. Thompson
LANDSCAPES AND
GEOMORPHOLOGY
Andrew Goudie and Heather Viles
LANGUAGES Stephen R. Anderson
LATE ANTIQUITY Gillian Clark
LAW Raymond Wacks
THE LAWS OF THERMODYNAMICS
Peter Atkins
LEADERSHIP Keith Grint
LEARNING Mark Haselgrove
LEIBNIZ Maria Rosa Antognazza

LEO TOLSTOY Liza Knapp
LIBERALISM Michael Freeden
LIGHT Ian Walmsley
LINCOLN Allen C. Guelzo
LINGUISTICS Peter Matthews
LITERARY THEORY Jonathan Culler
LOCKE John Dunn
LOGIC Graham Priest
LOVE Ronald de Sousa
MACHIAVELLI Quentin Skinner
MADNESS Andrew Scull
MAGIC Owen Davies
MAGNA CARTA Nicholas Vincent
MAGNETISM Stephen Blundell
MALTHUS Donald Winch
MAMMALS T. S. Kemp
MANAGEMENT John Hendry
MAO Delia Davin
MARINE BIOLOGY Philip V. Mladenov
THE MARQUIS DE SADE John Phillips
MARTIN LUTHER Scott H. Hendrix
MARTYRDOM Jolyon Mitchell
MARX Peter Singer
MATERIALS Christopher Hall
MATHEMATICAL FINANCE Mark H. A. Davis
MATHEMATICS Timothy Gowers
MATTER Geoff Cottrell
THE MEANING OF LIFE Terry Eagleton
MEASUREMENT David Hand
MEDICAL ETHICS Michael Dunn and Tony Hope
MEDICAL LAW Charles Foster
MEDIEVAL BRITAIN John Gillingham and Ralph A. Griffiths
MEDIEVAL LITERATURE Elaine Treharne
MEDIEVAL PHILOSOPHY John Marenbon
MEMORY Jonathan K. Foster
METAPHYSICS Stephen Mumford
METHODISM William J. Abraham
THE MEXICAN REVOLUTION Alan Knight
MICHAEL FARADAY Frank A. J. L. James
MICROBIOLOGY Nicholas P. Money
MICROECONOMICS Avinash Dixit
MICROSCOPY Terence Allen
THE MIDDLE AGES Miri Rubin
MILITARY JUSTICE Eugene R. Fidell
MILITARY STRATEGY Antulio J. Echevarria II
MINERALS David Vaughan
MIRACLES Yujin Nagasawa
MODERN ARCHITECTURE Adam Sharr
MODERN ART David Cottington
MODERN CHINA Rana Mitter
MODERN DRAMA Kirsten E. Shepherd-Barr
MODERN FRANCE Vanessa R. Schwartz
MODERN INDIA Craig Jeffrey
MODERN IRELAND Senia Pašeta
MODERN ITALY Anna Cento Bull
MODERN JAPAN Christopher Goto-Jones
MODERN LATIN AMERICAN LITERATURE Roberto González Echevarría
MODERN WAR Richard English
MODERNISM Christopher Butler
MOLECULAR BIOLOGY Aysha Divan and Janice A. Royds
MOLECULES Philip Ball
MONASTICISM Stephen J. Davis
THE MONGOLS Morris Rossabi
MOONS David A. Rothery
MORMONISM Richard Lyman Bushman
MOUNTAINS Martin F. Price
MUHAMMAD Jonathan A. C. Brown
MULTICULTURALISM Ali Rattansi
MULTILINGUALISM John C. Maher
MUSIC Nicholas Cook
MYTH Robert A. Segal
NAPOLEON David Bell
THE NAPOLEONIC WARS Mike Rapport
NATIONALISM Steven Grosby
NATIVE AMERICAN LITERATURE Sean Teuton
NAVIGATION Jim Bennett
NAZI GERMANY Jane Caplan
NELSON MANDELA Elleke Boehmer
NEOLIBERALISM Manfred Steger and Ravi Roy
NETWORKS Guido Caldarelli and Michele Catanzaro

THE NEW TESTAMENT Luke Timothy Johnson
THE NEW TESTAMENT AS LITERATURE Kyle Keefer
NEWTON Robert Iliffe
NIELS BOHR J. L. Heilbron
NIETZSCHE Michael Tanner
NINETEENTH–CENTURY BRITAIN Christopher Harvie and H. C. G. Matthew
THE NORMAN CONQUEST George Garnett
NORTH AMERICAN INDIANS Theda Perdue and Michael D. Green
NORTHERN IRELAND Marc Mulholland
NOTHING Frank Close
NUCLEAR PHYSICS Frank Close
NUCLEAR POWER Maxwell Irvine
NUCLEAR WEAPONS Joseph M. Siracusa
NUMBER THEORY Robin Wilson
NUMBERS Peter M. Higgins
NUTRITION David A. Bender
OBJECTIVITY Stephen Gaukroger
OCEANS Dorrik Stow
THE OLD TESTAMENT Michael D. Coogan
THE ORCHESTRA D. Kern Holoman
ORGANIC CHEMISTRY Graham Patrick
ORGANIZATIONS Mary Jo Hatch
ORGANIZED CRIME Georgios A. Antonopoulos and Georgios Papanicolaou
ORTHODOX CHRISTIANITY A. Edward Siecienski
PAGANISM Owen Davies
PAIN Rob Boddice
THE PALESTINIAN-ISRAELI CONFLICT Martin Bunton
PANDEMICS Christian W. McMillen
PARTICLE PHYSICS Frank Close
PAUL E. P. Sanders
PEACE Oliver P. Richmond
PENTECOSTALISM William K. Kay
PERCEPTION Brian Rogers
THE PERIODIC TABLE Eric R. Scerri
PHILOSOPHY Edward Craig
PHILOSOPHY IN THE ISLAMIC WORLD Peter Adamson
PHILOSOPHY OF BIOLOGY Samir Okasha
PHILOSOPHY OF LAW Raymond Wacks
PHILOSOPHY OF SCIENCE Samir Okasha
PHILOSOPHY OF RELIGION Tim Bayne
PHOTOGRAPHY Steve Edwards
PHYSICAL CHEMISTRY Peter Atkins
PHYSICS Sidney Perkowitz
PILGRIMAGE Ian Reader
PLAGUE Paul Slack
PLANETS David A. Rothery
PLANTS Timothy Walker
PLATE TECTONICS Peter Molnar
PLATO Julia Annas
POETRY Bernard O'Donoghue
POLITICAL PHILOSOPHY David Miller
POLITICS Kenneth Minogue
POPULISM Cas Mudde and Cristóbal Rovira Kaltwasser
POSTCOLONIALISM Robert Young
POSTMODERNISM Christopher Butler
POSTSTRUCTURALISM Catherine Belsey
POVERTY Philip N. Jefferson
PREHISTORY Chris Gosden
PRESOCRATIC PHILOSOPHY Catherine Osborne
PRIVACY Raymond Wacks
PROBABILITY John Haigh
PROGRESSIVISM Walter Nugent
PROHIBITION W. J. Rorabaugh
PROJECTS Andrew Davies
PROTESTANTISM Mark A. Noll
PSYCHIATRY Tom Burns
PSYCHOANALYSIS Daniel Pick
PSYCHOLOGY Gillian Butler and Freda McManus
PSYCHOLOGY OF MUSIC Elizabeth Hellmuth Margulis
PSYCHOPATHY Essi Viding
PSYCHOTHERAPY Tom Burns and Eva Burns-Lundgren
PUBLIC ADMINISTRATION Stella Z. Theodoulou and Ravi K. Roy
PUBLIC HEALTH Virginia Berridge

PURITANISM Francis J. Bremer
THE QUAKERS Pink Dandelion
QUANTUM THEORY
John Polkinghorne
RACISM Ali Rattansi
RADIOACTIVITY Claudio Tuniz
RASTAFARI Ennis B. Edmonds
READING Belinda Jack
THE REAGAN REVOLUTION Gil Troy
REALITY Jan Westerhoff
RECONSTRUCTION Allen. C. Guelzo
THE REFORMATION Peter Marshall
RELATIVITY Russell Stannard
RELIGION IN AMERICA Timothy Beal
THE RENAISSANCE Jerry Brotton
RENAISSANCE ART
Geraldine A. Johnson
RENEWABLE ENERGY Nick Jelley
REPTILES T. S. Kemp
REVOLUTIONS Jack A. Goldstone
RHETORIC Richard Toye
RISK Baruch Fischhoff and John Kadvany
RITUAL Barry Stephenson
RIVERS Nick Middleton
ROBOTICS Alan Winfield
ROCKS Jan Zalasiewicz
ROMAN BRITAIN Peter Salway
THE ROMAN EMPIRE
Christopher Kelly
THE ROMAN REPUBLIC
David M. Gwynn
ROMANTICISM Michael Ferber
ROUSSEAU Robert Wokler
RUSSELL A. C. Grayling
RUSSIAN HISTORY Geoffrey Hosking
RUSSIAN LITERATURE Catriona Kelly
THE RUSSIAN REVOLUTION
S. A. Smith
THE SAINTS Simon Yarrow
SAVANNAS Peter A. Furley
SCEPTICISM Duncan Pritchard
SCHIZOPHRENIA Chris Frith and
Eve Johnstone
SCHOPENHAUER
Christopher Janaway
SCIENCE AND RELIGION
Thomas Dixon
SCIENCE FICTION David Seed
THE SCIENTIFIC REVOLUTION
Lawrence M. Principe
SCOTLAND Rab Houston
SECULARISM Andrew Copson
SEXUAL SELECTION Marlene Zuk and
Leigh W. Simmons
SEXUALITY Véronique Mottier
SHAKESPEARE'S COMEDIES
Bart van Es
SHAKESPEARE'S SONNETS AND
POEMS Jonathan F. S. Post
SHAKESPEARE'S TRAGEDIES
Stanley Wells
SIKHISM Eleanor Nesbitt
THE SILK ROAD James A. Millward
SLANG Jonathon Green
SLEEP Steven W. Lockley and
Russell G. Foster
SOCIAL AND CULTURAL
ANTHROPOLOGY
John Monaghan and Peter Just
SOCIAL PSYCHOLOGY Richard J. Crisp
SOCIAL WORK Sally Holland and
Jonathan Scourfield
SOCIALISM Michael Newman
SOCIOLINGUISTICS John Edwards
SOCIOLOGY Steve Bruce
SOCRATES C. C. W. Taylor
SOUND Mike Goldsmith
SOUTHEAST ASIA James R. Rush
THE SOVIET UNION Stephen Lovell
THE SPANISH CIVIL WAR
Helen Graham
SPANISH LITERATURE Jo Labanyi
SPINOZA Roger Scruton
SPIRITUALITY Philip Sheldrake
SPORT Mike Cronin
STARS Andrew King
STATISTICS David J. Hand
STEM CELLS Jonathan Slack
STOICISM Brad Inwood
STRUCTURAL ENGINEERING
David Blockley
STUART BRITAIN John Morrill
SUPERCONDUCTIVITY
Stephen Blundell
SUPERSTITION Stuart Vyse
SYMMETRY Ian Stewart
SYNAESTHESIA Julia Simner
SYNTHETIC BIOLOGY Jamie A. Davies
SYSTEMS BIOLOGY Eberhard O. Voit
TAXATION Stephen Smith

TEETH Peter S. Ungar
TELESCOPES Geoff Cottrell
TERRORISM Charles Townshend
THEATRE Marvin Carlson
THEOLOGY David F. Ford
THINKING AND REASONING Jonathan St B. T. Evans
THOMAS AQUINAS Fergus Kerr
THOUGHT Tim Bayne
TIBETAN BUDDHISM Matthew T. Kapstein
TIDES David George Bowers and Emyr Martyn Roberts
TOCQUEVILLE Harvey C. Mansfield
TOPOLOGY Richard Earl
TRAGEDY Adrian Poole
TRANSLATION Matthew Reynolds
THE TREATY OF VERSAILLES Michael S. Neiberg
TRIGONOMETRY Glen Van Brummelen
THE TROJAN WAR Eric H. Cline
TRUST Katherine Hawley
THE TUDORS John Guy
TWENTIETH-CENTURY BRITAIN Kenneth O. Morgan
TYPOGRAPHY Paul Luna
THE UNITED NATIONS Jussi M. Hanhimäki
UNIVERSITIES AND COLLEGES David Palfreyman and Paul Temple
THE U.S. CONGRESS Donald A. Ritchie
THE U.S. CONSTITUTION David J. Bodenhamer
THE U.S. SUPREME COURT Linda Greenhouse
UTILITARIANISM Katarzyna de Lazari-Radek and Peter Singer
UTOPIANISM Lyman Tower Sargent
VETERINARY SCIENCE James Yeates
THE VIKINGS Julian D. Richards
VIRUSES Dorothy H. Crawford
VOLTAIRE Nicholas Cronk
WAR AND TECHNOLOGY Alex Roland
WATER John Finney
WAVES Mike Goldsmith
WEATHER Storm Dunlop
THE WELFARE STATE David Garland
WILLIAM SHAKESPEARE Stanley Wells
WITCHCRAFT Malcolm Gaskill
WITTGENSTEIN A. C. Grayling
WORK Stephen Fineman
WORLD MUSIC Philip Bohlman
THE WORLD TRADE ORGANIZATION Amrita Narlikar
WORLD WAR II Gerhard L. Weinberg
WRITING AND SCRIPT Andrew Robinson
ZIONISM Michael Stanislawski

Available soon:

SMELL Matthew Cobb
THE SUN Philip Judge
ENZYMES Paul Engel
FIRE Andrew C. Scott
ECOLOGY Jaboury Ghazoul

For more information visit our web site
www.oup.com/vsi/

Robin Wilson

NUMBER THEORY

A Very Short Introduction

Great Clarendon Street, Oxford OX2 6DP,
United Kingdom

Oxford University Press is a department of the University of Oxford. It furthers the University's objective of excellence in research, scholarship, and education by publishing worldwide. Oxford is a registered trade mark of Oxford University Press in the UK and in certain other countries

First Edition published in 2020

Published in the United States of America by Oxford University Press
198 Madison Avenue, New York, NY 10016, United States of America

British Library Cataloguing in Publication Data
Data available

Library of Congress Control Number: 2020932768

ISBN 978–0–19–879809–5

Printed and bound by
CPI Group (UK) Ltd, Croydon, CR0 4YY

Contents

List of illustrations

Chapter 1
What is number theory?

Consider the following questions:

In which years does February have five Sundays?

What is special about the number 4,294,967,297*?*

How many right-angled triangles with whole-number sides have a side of length 29?

Are any of the numbers 11, 111, 1111, 11111, ... *perfect squares?*

I have some eggs. When arranged in rows of 3 *there are* 2 *left over, in rows of* 5 *there are* 3 *left over, and in rows of* 7 *there are* 2 *left over. How many eggs have I altogether?*

Can one construct a regular polygon with 100 *sides if measuring is forbidden?*

How many shuffles are needed to restore the order of the cards in a pack with two Jokers?

If I can buy partridges for 3 *cents, pigeons for* 2 *cents, and* 2 *sparrows for a cent, and if I spend* 30 *cents on buying* 30 *birds, how many birds of each kind must I buy?*

How do prime numbers help to keep our credit cards secure?

What is the Riemann hypothesis, and how can I earn a million dollars?

As you'll discover, these are all questions in *number theory*, the branch of mathematics that's primarily concerned with our *counting numbers*, 1, 2, 3, . . . , and we'll meet all of these questions again later. Of particular importance to us will be the *prime numbers*, the 'building blocks' of our number system: these are numbers such as 19, 199, and 1999 whose only factors are themselves and 1, unlike 99 which is 9×11 and 999 which is 27×37. Much of this book is concerned with exploring their properties.

Number theory is an old subject, dating back over two millennia to the Ancient Greeks. The Greek word ἀριθμὸς (arithmos) means 'number', and for the Pythagoreans of the 6th century BC 'arithmetic' originally referred to calculating with whole numbers, and by extension to what we now call number theory—in fact, until fairly recently the subject was sometimes referred to as 'the higher arithmetic'. Three centuries later, Euclid of Alexandria discussed arithmetic and number theory in Books VII, VIII, and IX of his celebrated work, the *Elements*, and proved in particular that the list of prime numbers is never-ending. Then, possibly around AD 250, Diophantus, another inhabitant of Alexandria, wrote a classic text called *Arithmetica* which contained many questions with whole number solutions.

After the Greeks, there was little interest in number theory for over one thousand years until the pioneering insights of the 17th-century French lawyer and mathematician Pierre de Fermat, after whom 'Fermat's last theorem', one of the most celebrated challenges of number theory, is named. Fermat's work was developed by the 18th-century Swiss polymath Leonhard Euler, who solved several problems that Fermat had been unable to crack, and also by Joseph-Louis Lagrange in Berlin and Adrien-Marie Legendre in Paris. In 1793 the German prodigy Carl Friedrich Gauss constructed by hand a list of all the prime numbers up to three million when he was aged just 15, and shortly afterwards wrote a groundbreaking text entitled *Disquisitiones*

Arithmeticae (Investigations into Arithmetic) whose publication in 1801 revolutionized the subject. Sometimes described as the 'Prince of Mathematics', Gauss asserted that

> Mathematics is the queen of the sciences, and number theory is the queen of mathematics.

The names of these trailblazers will reappear throughout this book (see Figure 1).

1. From left to right; Euclid, Pierre de Fermat, Leonhard Euler, and Carl Friedrich Gauss.

More recently, the subject's scope has broadened greatly to include many other topics, several of which feature in this book. In particular, there have been some spectacular developments, such as Andrew Wiles's proof of Fermat's last theorem (which had remained unproved for over 350 years) and some exciting new results on the way that prime numbers are distributed.

Number theory has long been thought of as one of the most 'beautiful' areas of mathematics, exhibiting great charm and elegance: prime numbers even arise in nature, as we'll see. It's also one of the most tantalizing of subjects, in that several of its challenges are so easy to state that anyone can understand them—and yet, despite valiant attempts by many people over hundreds of years, they've never been solved. But the subject has also recently become of great practical importance—in the area of cryptography. Indeed, somewhat surprisingly, much secret information, including the security of your credit cards, depends on a result from number theory that dates back to the 18th century.

In this chapter I'll lay the groundwork for our later explorations, by introducing several types of number that you'll meet again and posing several questions. Some of these questions are easy to answer, whereas others are harder but are solved in subsequent chapters, and a few are notorious problems for which no answer has yet been found. For the moment I won't reveal which questions fall into which category, because you may like to think about them first. Their answers (where known) are summarized at the end of this book, in Chapter 9.

Integers

This book is about the *integers* (or whole numbers),

$$\ldots, -4, -3, -2, -1, 0, 1, 2, 3, 4, 5, \ldots.$$

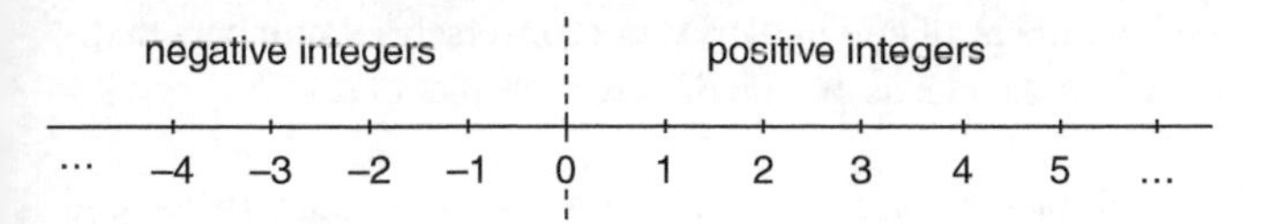

2. The integers.

These include the counting numbers or *positive integers* (1, 2, 3, 4, 5, . . .), the *negative integers* (. . . , −4, −3, −2, −1), and the number 0 (see Figure 2).

We can also split the collection of integers into the *even numbers*

$$\ldots, -6, -4, -2, 0, 2, 4, 6, 8, \ldots$$

and the *odd numbers*

$$\ldots, -5, -3, -1, 1, 3, 5, 7, \ldots .$$

Every even number is twice another integer—that is, it has the form $2n$, where n is an integer: for example,

$$12 = 2 \times 6, \quad \text{where } n = 6.$$

Similarly, every odd number is one more than twice another integer—that is, it has the form $2n + 1$, where n is an integer: for example,

$$13 = (2 \times 6) + 1, \quad \text{where again } n = 6.$$

The *multiples* of a given integer n are those numbers that leave no remainder when divided by n: for example,

the positive multiples of 10 are $10, 20, 30, 40, 50, \ldots .$

These numbers all end in 0, and conversely all numbers that end with 0 (such as 70) are multiples of 10. Similarly,

the positive multiples of 5 are $5, 10, 15, 20, 25, \ldots .$

These numbers all end in 0 or 5, and conversely all numbers that end in 0 or 5 (such as 40 and 65) are multiples of 5.

The multiples of 2 are the even numbers—those numbers that end in 2, 4, 6, 8, or 0. But what can we say about the multiples of other numbers? For example:

How can we recognize whether a given number, such as 12,345,678, *is a multiple of* 8*? or of* 9*? or of* 11*? or of* 88*?*

I'll answer these questions in Chapter 2, where we'll explore multiples in greater detail.

In number theory the single word 'number' generally refers to a positive integer, and we shall follow this convention unless otherwise stated.

Squares and cubes

The Pythagoreans seem to have been particularly interested in perfect squares, which they depicted geometrically by square patterns of dots, as in Figure 3.

3. The first four non-zero squares.

A *square* (or *perfect square*) has the form $n^2 = n \times n$, where n is an integer: for example, 144 is a square because $144 = 12^2$ or $(-12)^2$, and 0 is a square because $0 = 0^2$. All the non-zero squares are positive integers, the first ten being

$$1^2 = 1, \quad 2^2 = 4, \quad 3^2 = 9, \quad 4^2 = 16, \quad 5^2 = 25, \quad 6^2 = 36,$$
$$7^2 = 49, \quad 8^2 = 64, \quad 9^2 = 81, \quad \text{and} \quad 10^2 = 100.$$

These squares all end in 1, 4, 5, 6, 9, or 0, and we might ask:

Do any squares end in 2, 3, 7, *or* 8*?*

We also notice that each of these squares is

either a multiple of 4: for example, $36 = 4 \times 9$,

or one more than a multiple of 4: for example, $49 = (4 \times 12) + 1$,

and we might ask whether this is always true:

Must all squares be of the form $4n$ *or* $4n + 1$, *where* n *is an integer?*

The Pythagoreans also apparently observed such results as

$$1 + 3 + 5 + 7 = 16 \quad \text{and} \quad 1 + 3 + 5 + 7 + 9 + 11 + 13 = 49.$$

Because 16 and 49 are both squares, we might ask:

Must the sum of the first few odd numbers, 1, 3, 5, 7, ..., *always be a square?*

What happens if we add two squares together? The number $13 = 3^2 + 2^2$ can be written as the sum of two squares, as can the number $17 = 4^2 + 1^2$. But the numbers 12, 14, and 15 cannot be written in this way, and we may ask:

Which numbers can be written as the sum of two squares?

However, the numbers $12 = 2^2 + 2^2 + 2^2$ and $14 = 3^2 + 2^2 + 1^2$ can be written as the sum of *three* squares, and the number $15 = 3^2 + 2^2 + 1^2 + 1^2$ can be written as the sum of *four* squares, and we can ask such questions as:

Can 9999 *be written as the sum of two squares? or of three squares? or of four squares?*

Squares also arise in the geometry of right-angled triangles. By the Pythagorean theorem, the lengths a, b, c of the sides of a right-angled triangle satisfy the equation $a^2 + b^2 = c^2$ (see Figure 4): for example,

$$3^2 + 4^2 = 5^2 \quad \text{and} \quad 5^2 + 12^2 = 13^2,$$

and we might ask:

Which other right-angled triangles have integer-length sides?

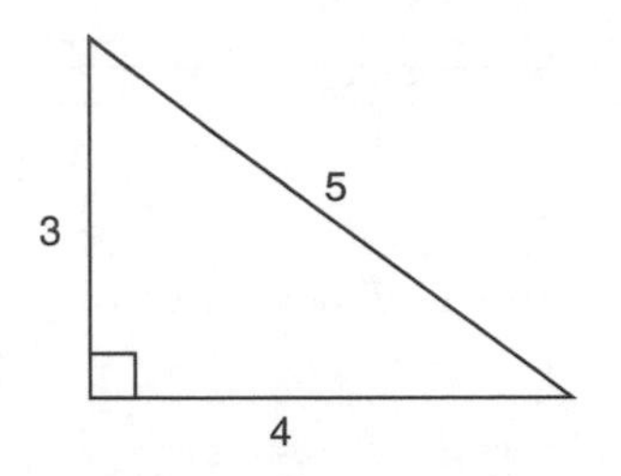

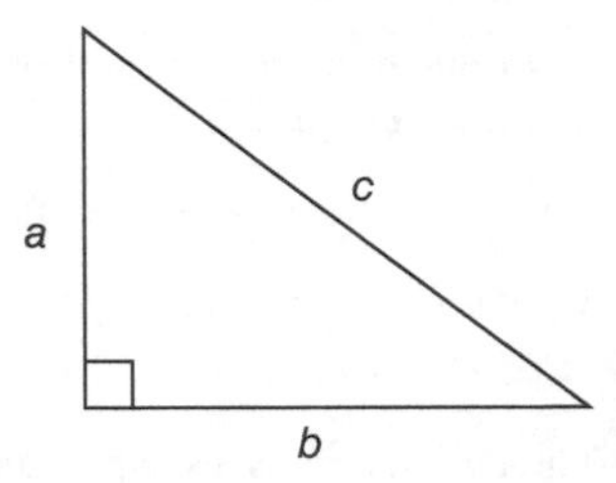

4. Right-angled triangles.

Let's now turn our attention to cubes.

A *cube* (or *perfect cube*) has the form $n^3 = n \times n \times n$, where n is an integer: for example, 343 and -216 are cubes because $343 = 7^3$ and $-216 = (-6)^3$. Cubes can be positive, negative, or zero, and the first ten positive cubes are

$$1^3 = 1, \quad 2^3 = 8, \quad 3^3 = 27, \quad 4^3 = 64, \quad 5^3 = 125,$$
$$6^3 = 216, \quad 7^3 = 343, \quad 8^3 = 512, \quad 9^3 = 729,$$
$$10^3 = 1000.$$

Each of these cubes is

either a multiple of 9: for example, $27 = 9 \times 3$,

or one more than a multiple of 9: for example, $64 = (9 \times 7) + 1$,

or eight more than a multiple of 9: for example, $125 = (9 \times 13) + 8$,

and we might ask:

Must all cubes be of the form $9n$, $9n + 1$, or $9n + 8$, where n is an integer?

An unexpected link between squares and cubes is

$$1^3 + 2^3 + 3^3 + 4^3 = 100 = 10^2$$
$$1^3 + 2^3 + 3^3 + 4^3 + 5^3 = 225 = 15^2,$$

and we might ask:

Must the sum of the first few cubes $1^3, 2^3, 3^3, \ldots$ *always be a square?*

Above we saw that the sum of two squares can be another square: for example, $3^2 + 4^2 = 5^2$. We may ask whether there's a similar statement for cubes:

Are there any integers a, b, c for which $a^3 + b^3 = c^3$*?*

Just as we can write numbers as the sum of squares, so we can also write them as the sum of cubes—for example:

$$200 = 125 + 64 + 8 + 1 + 1 + 1 = 5^3 + 4^3 + 2^3 + 1^3 + 1^3 + 1^3.$$

So we might ask:

Can every number be written as the sum of six cubes?

I'll answer these questions in Chapters 2 and 5 where we'll explore squares and cubes in greater detail

Perfect numbers

The *factors*, or *divisors*, of a given number are the positive integers that divide exactly into it, leaving no remainder: for example, the factors of 10 are 1, 2, 5, and 10. A factor that's not equal to the number itself is a *proper factor*: the proper factors of 10 are 1, 2, and 5.

In Book IX of his *Elements* Euclid discussed perfect numbers, which were believed to have mystic or religious significance. A *perfect number* is a number whose proper factors add up to the original number: for example,

6 is perfect, because its proper factors are 1, 2, and 3, which add up to 6;

28 is perfect, because its proper factors are 1, 2, 4, 7, and 14, which add up to 28.

The first four perfect numbers, already known to the Ancient Greeks, are 6, 28, 496, and 8128, and we might ask:

What is the next perfect number after 8128?

and, more generally,

Is there a formula for producing perfect numbers?

We'll explore perfect numbers in Chapter 3.

Prime numbers

As you saw earlier, a *prime number* is a number that has no factors other than itself and 1: for example, 13 and 17 are prime numbers, whereas $15 = 3 \times 5$ is not. The first fifteen primes are

$$2, 3, 5, 7, 11, 13, 17, 19, 23, 29, 31, 37, 41, 43, \text{ and } 47.$$

A number that is not prime (such as 14, 15, or 16) is called *composite.*

The number 1 is regarded as neither prime nor composite. We'll explain why in Chapter 3, where we explore prime numbers in greater detail.

Prime numbers lie at the heart of number theory because they're the 'building blocks', or 'atoms', of our counting system, in the sense that every number that's greater than 1 can be obtained by multiplying primes together: for example,

$$90 = 2 \times 3 \times 3 \times 5 \quad \text{and} \quad 91 = 7 \times 13.$$

In some cases we can answer difficult questions about numbers in general by first answering them for primes and then combining the results.

From the above list we see that 2 and 3 seem to be the only prime numbers that differ by 1, but that several pairs of primes differ by 2: some examples are

$$3 \text{ and } 5, \quad 5 \text{ and } 7, \quad 29 \text{ and } 31, \quad \text{and} \quad 41 \text{ and } 43.$$

Such pairs are called *twin primes*, and larger examples include 101 and 103, 2027 and 2029, and 9,999,971 and 9,999,973. Knowing that the list of primes is never-ending, we may likewise ask:

Does the list of twin primes go on for ever?

On the other hand, we sometimes find large gaps between successive prime numbers; for example, the prime numbers 23 and 29 are separated by the five composite numbers, 24, 25, 26, 27, and 28, and the primes 113 and 127 are separated by the thirteen consecutive composite numbers from 114 to 126. But how large can these gaps be? For example:

Is there a string of 1000 *consecutive composite numbers?*

Another question arises when we add prime numbers. Noticing that

$$4 = 2 + 2, \quad 12 = 7 + 5, \quad \text{and} \quad 20 = 13 + 7 \text{ or } 17 + 3,$$

we might ask:

Can every even number be written as the sum of two primes?

Several prime numbers can be written as the sum of two squares: for example,

$$5 = 4 + 1, \quad 13 = 9 + 4, \quad 29 = 25 + 4, \quad \text{and} \quad 41 = 25 + 16.$$

But some other primes, such as 11, 23, and 47, cannot be written like this, and we might ask:

Which prime numbers can be written as the sum of two squares?

We may also notice that some prime numbers are *one less than a power of* 2: for example,

$$3 = 2^2 - 1, \quad 7 = 2^3 - 1, \quad 31 = 2^5 - 1, \quad \text{and} \quad 127 = 2^7 - 1.$$

But none of the following numbers of this type is prime:

$$15 = 2^4 - 1, \quad 63 = 2^6 - 1, \quad 255 = 2^8 - 1, \quad 511 = 2^9 - 1,$$
$$1023 = 2^{10} - 1.$$

Noticing that the exponents (2, 3, 5, and 7) in the first list are all prime, whereas those in the second list (4, 6, 8, 9, and 10) are all composite, we might ask:

Is the number $2^n - 1$ always prime when n is prime, and always composite when n is composite?

Prime numbers of the form $2^n - 1$ are called *Mersenne primes* after the 17th-century French mathematician and friar Marin Mersenne, who explored their properties. They arise in connection with perfect numbers, and with the search for large primes, as we'll see in Chapter 3.

Likewise, some prime numbers are *one more than a power of* 2. For example, Pierre de Fermat considered numbers of the form $2^n + 1$, where n is itself a power of 2, and when $n = 1, 2, 4, 8,$ and 16, he obtained the numbers

$$2^1 + 1 = 2 + 1 = 3, \quad 2^2 + 1 = 4 + 1 = 5,$$
$$2^4 + 1 = 16 + 1 = 17, \quad 2^8 + 1 = 256 + 1 = 257,$$
$$2^{16} + 1 = 65,536 + 1 = 65,537.$$

Recognizing that these five numbers are all prime, Fermat tried (but failed) to prove that this was always the case, and so we may ask:

Are all numbers of this form prime?

These numbers are now called *Fermat numbers*, and we'll investigate them in Chapter 3, where we'll see how they arise unexpectedly in connection with an ancient problem in geometry.

Before leaving prime numbers, we notice that every prime number (other than 2), being an odd number, must be either one more, or three more, than a multiple of 4—that is, it's of the form $4n + 1$ or $4n + 3$, for some integer n. Examples of primes of the first type are

$$5 = (4 \times 1) + 1, \quad 13 = (4 \times 3) + 1, \quad 17 = (4 \times 4) + 1,$$
$$29 = (4 \times 7) + 1,$$

and of the second type are

$$7 = (4 \times 1) + 3, \quad 11 = (4 \times 2) + 3, \quad 19 = (4 \times 4) + 3,$$
$$23 = (4 \times 5) + 3.$$

Do these lists go on for ever?—that is, we may ask:

Are there infinitely many primes of the form $4n + 1$? or of the form $4n + 3$?

We may also ask the following related question:

Are there infinitely many primes with final digit 9?

We'll explore such questions in Chapter 7.

This introductory chapter was designed to give you some idea of what to expect in the coming chapters, and some intimation of the delights that await you as we explore an area of study that has fascinated amateurs and professionals alike for thousands of years. Number theory is now a massive subject and many important topics have had to be omitted from these pages, but I hope that my selection will give you some idea of the wide-ranging aspects of number theory as it arose historically and as it is still practised today.

Chapter 2
Multiplying and dividing

Much of number theory is concerned with multiplying and dividing whole numbers, and in this chapter we'll explore their multiples and divisors (or factors). After presenting the division rule and Euclid's algorithm for finding the greatest common divisor of two numbers, we'll explore some properties of squares and cubes, present some quick tests for determining which numbers can be divided evenly by certain given numbers (such as 4, 9, and 11), and conclude with the ancient method of 'casting out nines'.

Multiples and divisors

Given two integers a and b, we say that *b is a multiple of a* if there's an integer x with $a \times x = b$: for example, 18 is a multiple of 3 because $3 \times 6 = 18$; here, $x = 6$. In this case, we also say that a is a *divisor* or a *factor* of b, and that *a divides b*, and b is *divisible* by a, so 3 is a divisor or factor of 18, 3 divides 18, and 18 is divisible by 3 (see Figure 5). All of these terms are in common use and we'll use them interchangeably.

As further examples, we see that the first five positive multiples of 100 are

100, 200, 300, 400, and 500,

and that the positive divisors of 100 are

1, 2, 4, 5, 10, 20, 25, 50, and 100.

Also, the number 1 divides every positive integer.

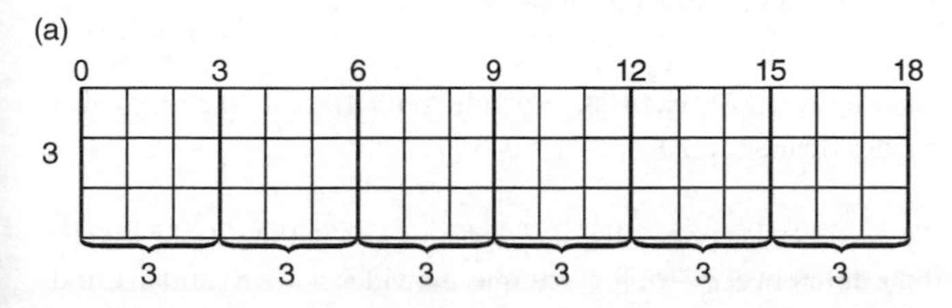

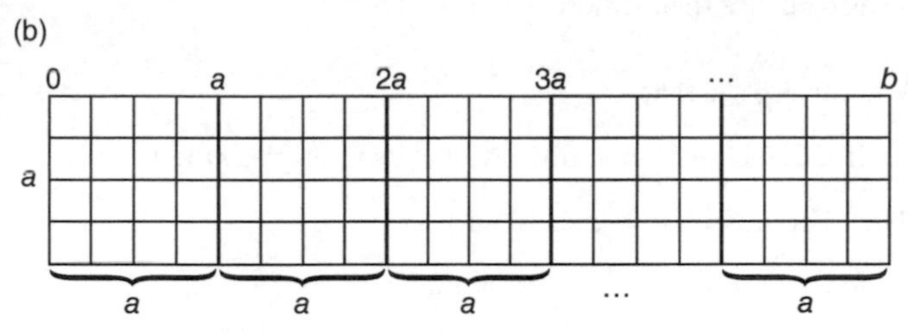

5. 18 is a multiple of 3, and 3 is a divisor of 18;
***b* is a multiple of *a*, and *a* is a divisor of *b*.**

A basic result on multiples and divisors is:

> If a and b are both multiples of a number d, then so is their sum $a + b$,

or, in terms of divisors,

> If d divides both a and b, then d also divides their sum $a + b$.

This is because if $a = d \times x$ and $b = d \times y$, for some numbers x and y, then

$$a + b = (d \times x) + (d \times y) = d \times (x + y)$$

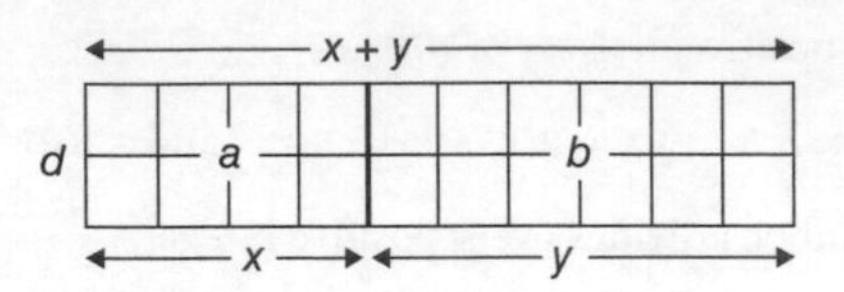

6. If d divides a and b, then d also divides $a + b$.

(see Figure 6). For example, 5 divides both 40 and 30, and so also divides their sum, 70.

We can likewise prove that if d divides a and b, then d also divides their difference, $a - b$: for example, 5 divides both 40 and 30, and so also divides their difference, 10.

We can also see that

> If d divides a, then d divides all of its multiples $m \times a$.

This is because, if $a = d \times x$, then

$$m \times a = m \times (d \times x) = d \times (m \times x):$$

for example, 5 divides 40, and so divides all of its multiples, such as 160.

Combining this with the above result about the sum, we deduce that:

> If d divides both a and b, then d divides all numbers of the form $(m \times a) + (n \times b)$, for any integers m and n.

This is because d divides the multiples $m \times a$ and $n \times b$, and so divides their sum: we'll call these numbers 'combinations' of a and b. For example, because 5 divides both 40 and 30, it also divides their multiples $4 \times 40 = 160$ and $3 \times 30 = 90$, and so divides the sum of these, the combination $250 = (4 \times 40) + (3 \times 30)$. We notice that the special cases $m = 1, n = 1$, and $m = 1, n = -1$, give us the above statements on the sum $a + b$ and difference $a - b$.

For a change of pace, we'll end this section with a puzzle that involves divisors:

A census-taker visits a mathematical family at their home and the following conversation takes place:

Census-taker: *How old are your three children?*

Parent: *The product of their ages is 36 and the sum is our house number.*

Census-taker: *I need more information. Are your two youngest children the same age?*

Parent: *No.*

Census-taker: *Ah! Now I know their ages.*

How old were they?

How might we go about answering this, as there seems to be too little information provided? To do so, let's look first at the possible ways of writing 36 as the product of three numbers:

$$36 \times 1 \times 1, \quad 18 \times 2 \times 1, \quad 12 \times 3 \times 1, \quad 9 \times 4 \times 1,$$
$$9 \times 2 \times 2, \quad 6 \times 6 \times 1, \quad 6 \times 3 \times 2, \quad \text{and} \quad 4 \times 3 \times 3,$$

with sums of 38, 21, 16, 14, 13, 13, 11, and 10, respectively. The census-taker, seeing the house number, would know which of these was correct, unless the sum were the repeated number 13, in which case there are two possibilities. But because the two youngest children are not the same age, their ages cannot be 9, 2, and 2, and so must be 6, 6, and 1.

Least common multiple and greatest common divisor

In this section we'll investigate two important numbers associated with the numbers *a* and *b*.

The least common multiple

Let's look at two situations. The first involves two ancient calendars:

In the first millennium AD *the Mayans of Central America had two yearly calendars, one based on* 260 *days and the other on* 365 *days, which they then combined into a single 'calendar round' of* 18,980 *days* (= 52 *years*). *But where did this number come from?*

The second situation involves two gears (see Figure 7):

I have two rotating gears, with 90 *and* 54 *teeth. When do the starting positions of these gears align?*

The starting positions align whenever the number of teeth that have passed the starting position is simultaneously a multiple of 90 and a multiple of 54. So what are these multiples?

For the first gear, the first few multiples are

90, 180, **270**, 360, 450, **540**, and 630,

whereas for the second gear, they are

54, 108, 162, 216, **270**, 324, 378, 432, 486, **540**, and 594,

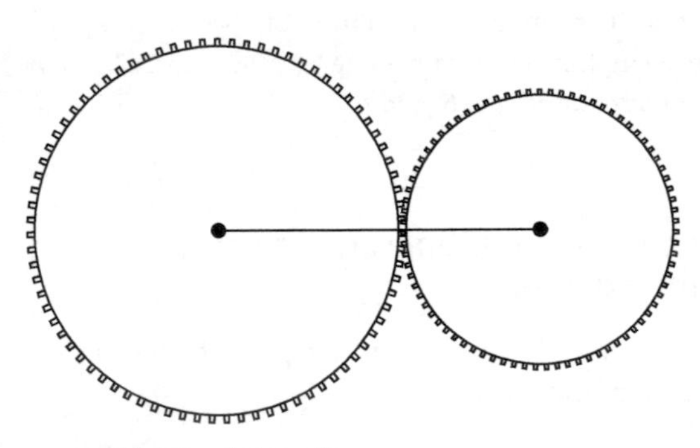

7. Two gears with 90 and 54 teeth.

and the multiples that they have in common are the numbers in bold type—that is, 270 and 540. The smaller of these is 270, and we say that 270 is the 'least common multiple' of 90 and 54. So the starting positions align whenever 270 teeth have passed—that is, after every three rotations of the first gear and every five rotations of the second gear.

In general, m is a *common multiple* of the integers a and b if m is a multiple of both a and b, and the *least common multiple* is the smallest positive common multiple. If m is the least common multiple of a and b, we write $m = \text{lcm}\,(a, b)$. In the above example,

$$\text{lcm}\,(90, 54) = 270,$$

and further examples are

$$\text{lcm}\,(15, 10) = 30, \quad \text{lcm}\,(16, 10) = 80,$$
$$\text{lcm}\,(20, 10) = 20.$$

For the Ancient Mayans, the two calendars came together after each period of $\text{lcm}\,(260, 365) = 18{,}980$ days.

Many people meet least common multiples for the first time when they learn how to add fractions. For example, to add the fractions 1/90 and 1/54 we first put them over a common denominator:

$$\frac{1}{90} + \frac{1}{54} = \frac{3}{270} + \frac{5}{270}.$$

We can now add these fractions directly to give 8/270, which then simplifies to 4/135. Here, the common denominator is the least common multiple, lcm (90, 54) = 270.

The greatest common divisor

Related to the least common multiple of two integers is their *greatest common divisor*.

The divisors of 90 are

$$1, 2, 3, 5, 6, 9, 10, 15, 18, 30, 45, \text{and } 90,$$

and those of 54 are

$$1, 2, 3, 6, 9, 18, 27, \text{and } 54,$$

so the ones that they have in common are 1, 2, 3, 6, 9, and 18. The largest of these is 18, so 18 is the 'greatest common divisor' of 90 and 54.

In general, d is a *common divisor* of the numbers a and b if d divides both a and b, and the *greatest common divisor* is the largest of these common divisors. If d is the greatest common divisor of a and b, we write $d = \gcd(a, b)$; it is sometimes called their *highest common factor* (written hcf). In the example above,

$$\gcd(90, 54) = 18,$$

and further examples are

$$\gcd(14, 10) = 2, \quad \gcd(17, 10) = 1, \quad \gcd(25, 10) = 5.$$

If $\gcd(a, b) = 1$, we say that a and b are *relatively prime* or *coprime*: for example, 17 and 10 have no positive factors in common, except 1, and so are coprime.

Surprisingly, these ideas even arise in nature—in the life cycles of certain insects. In North America three types of cicada (see Figure 8) have life cycles of 7, 13, and 17 years, all of which are prime numbers. Is this a coincidence? Cicadas stay underground for most of their lives, and then emerge all at once for an orgy of eating, chirping, mating, laying eggs, and then dying. But when they do appear they're vulnerable to predators (such as birds and certain wasps) with shorter life cycles of up to five years. If a cicada's life cycle were 12 years, or some other composite number, then the likelihood of being consumed by a predator would be greatly increased. But because they've evolved life cycles of 7, 13, and 17 years, and because these numbers are all coprime to 2, 3, 4, and 5, the cicadas can more easily avoid that unhappy fate.

8. A periodical cicada.

A fundamental property of the greatest common divisor d of two numbers a and b is that we can always write d as a combination of a and b. To see what is involved, consider the following simple problem involving American coinage:

Jack and Jill have a number of quarters (25 cents) and dimes (10 cents), and Jack wishes to pay Jill 5 cents. How can this be done?

One way is for Jack to give Jill 1 quarter and for Jill to give Jack 2 dimes. We can write this as

$$5 = (1 \times 25) + (-2 \times 10)\,.$$

Another way is for Jill to give Jack 1 quarter and for Jack to give Jill 3 dimes. We can write this as

$$5 = (-1 \times 25) + (3 \times 10)\,.$$

Taking $a = 25$ and $b = 10$ and noting that $\gcd(a, b) = 5$, we can generalize these observations as follows:

> If a and b are positive integers, and if $d = \gcd(a, b)$, then there are integers m and n for which $d = (m \times a) + (n \times b)$.

In particular, if a and b are coprime, then $\gcd(a, b) = 1$, and this result tells us that there are integers m and n for which

$$1 = (m \times a) + (n \times b)\,.$$

As above, the integers m and n can't both be positive, and there are many possible choices for them: for example, $\gcd(90, 54) = 18$ and

$$18 = (-1 \times 90) + (2 \times 54): \quad \text{here, } m = -1 \text{ and } n = 2$$
$$18 = (2 \times 90) + (-3 \times 54): \quad \text{here, } m = 2 \text{ and } n = -3.$$

We'll conclude this section with an interesting connection between the least common multiple and the greatest common divisor and of two numbers a and b: this is

$$\operatorname{lcm}(a, b) \times \gcd(a, b) = a \times b.$$

For example, if $a = 90$ and $b = 54$, then $\operatorname{lcm}(a, b) = 270$, $\gcd(a, b) = 18$, and

$$\operatorname{lcm}(a, b) \times \gcd(a, b) = 270 \times 18 = 4860 = 90 \times 54 = a \times b.$$

In Chapter 3 we'll explain why this happens.

Euclid's algorithm

How can we calculate the greatest common divisor of two given numbers?

In Book VII of his *Elements*, Euclid presented a method for doing so which depends on an elementary, yet fundamental, result that's known as the 'division rule'. This tells us that if we're given any integers a and b, then we can divide b by a to give an answer (called a 'quotient'), usually with a remainder: for example, if we divide 34 by 10 we get a quotient of 3 and a remainder of 4, because

$$34 = (3 \times 10) + 4.$$

We notice that the remainder is less than 10, the number that we divided by.

> *Division rule*: Given any positive integers a and b, there are unique numbers q (the *quotient*) and r (the *remainder*) with $b = (q \times a) + r$, where $0 \le r < a$.

This result is illustrated in Figure 9. As a special case, if b is a multiple of a, then the remainder r is 0 and the quotient q is b/a.

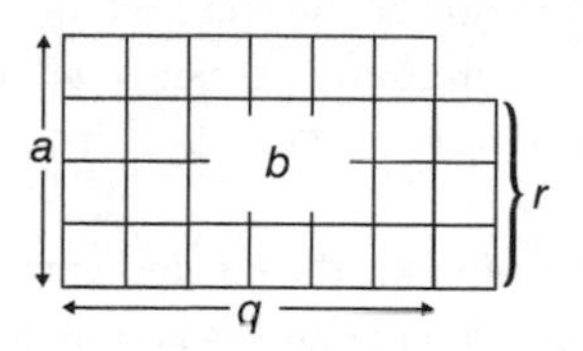

9. The division rule.

Some important special cases of the division rule, which we'll need later in this chapter, are as follows; they are illustrated in Figure 10:

If $a = 2$, then $r = 0$ or 1, so:

> Every integer b has the form $2q$ (the even integers) or $2q + 1$ (the odd integers).

If $a = 3$, then $r = 0, 1$, or 2, so:

> Every integer b has the form $3q$, $3q + 1$, or $3q + 2$.

If $a = 4$, then $r = 0, 1, 2$, or 3, so:

> Every integer b has the form $4q$, $4q + 1$, $4q + 2$, or $4q + 3$,

and so on.

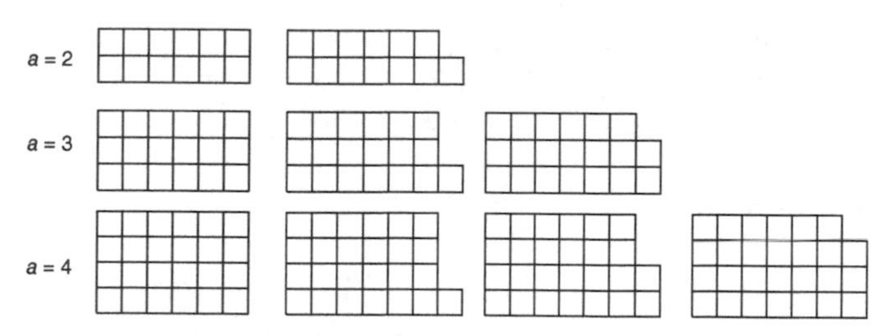

10. Special cases of the division rule.

We come now to Euclid's algorithm. An *algorithm* is a finite procedure for solving a problem, step by step. It's rather like a recipe in a cookery book or a set of road instructions for driving from one location to another. When we supply the appropriate input data (the ingredients or the two locations) and 'turn the handle', our output should be the required solution (such as a cake or a suitable route). Algorithms are named after the 9th-century Persian mathematician al-Khwārizmī.

Euclid's algorithm for finding the greatest common divisor of two numbers uses the division rule over and over again. The following example, where we show that gcd (57, 21) = 3, illustrates the method. Here, the numbers in bold type are the original numbers 57 and 21 and the remainders that arise in the successive divisions.

First, divide 57 by 21: $\mathbf{57} = (2 \times \mathbf{21}) + \mathbf{15}$
Next, divide 21 by 15: $\mathbf{21} = (1 \times \mathbf{15}) + \mathbf{6}$
Next, divide 15 by 6: $\mathbf{15} = (2 \times \mathbf{6}) + \mathbf{3}$
Finally, divide 6 by 3: $\mathbf{6} = (2 \times \mathbf{3}) + \mathbf{0}$.

We stop when we obtain a remainder of 0, and the greatest common divisor is then *the last non-zero remainder*, which here is 3. Figure 11 depicts this process.

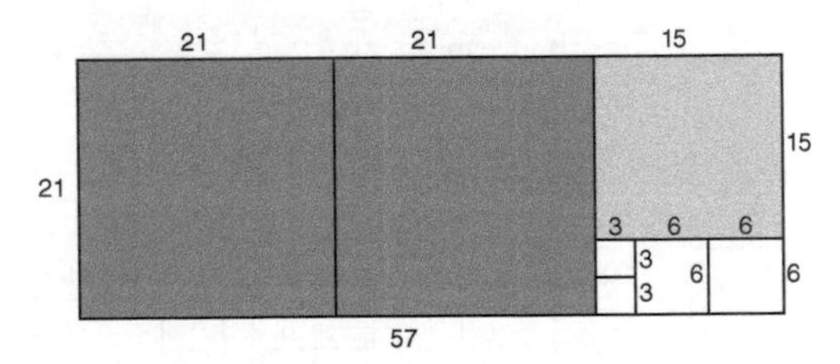

11. gcd (57, 21) = 3.

We can also use these same calculations to write gcd (57, 21) as a combination of 57 and 21, as we described earlier. To do so *we work upwards from the last-but-one equation*, substituting for the remainder at each step, as follows.

By the last-but-one equation, $\mathbf{3} = \mathbf{15} - (2 \times \mathbf{6})$.
But by the second equation, $\mathbf{6} = \mathbf{21} - (1 \times \mathbf{15})$,

and substituting this into the last-but-one equation gives

$$\mathbf{3} = \mathbf{15} - \{2 \times (\mathbf{21} - (1 \times \mathbf{15}))\},$$

which simplifies to $\mathbf{3} = (3 \times \mathbf{15}) - (2 \times \mathbf{21})$.

Finally, by the first application of the division rule,

$$\mathbf{15} = \mathbf{57} - (2 \times \mathbf{21}),$$

and substituting this into the previous equation gives

$$\mathbf{3} = 3 \times \{\mathbf{57} - (2 \times \mathbf{21})\} - (2 \times \mathbf{21}),$$

which simplifies to $\mathbf{3} = (3 \times \mathbf{57}) + (-8 \times \mathbf{21})$.

So $\gcd(57, 21) = 57m + 21n$, where $m = 3$ and $n = -8$.

In general, we can find the greatest common divisor of any two positive integers a and b in the same way—by using the division rule repeatedly to find each quotient and remainder in turn, until we get a remainder of 0. Then:

> The greatest common divisor gcd (a, b) is the last non-zero remainder.

We can then reverse the process to write gcd (a, b) as a combination of a and b. To do so, we start with the last-but-one equation and work upwards through the equations, as in the example we've just given.

How efficient is Euclid's algorithm? Sometimes Euclid's method works quickly and we find the highest common factor in a small number of steps. For example, to show that gcd (90, 54) = 18 requires just three steps:

$$\mathbf{90} = (1 \times \mathbf{54}) + \mathbf{36}$$
$$\mathbf{54} = (1 \times \mathbf{36}) + \mathbf{18}$$
$$\mathbf{36} = (2 \times \mathbf{18}) + \mathbf{0}.$$

So gcd (90, 54) = 18 (see Figure 12).

But sometimes it takes more steps—for example, when it is applied to successive numbers in the sequence

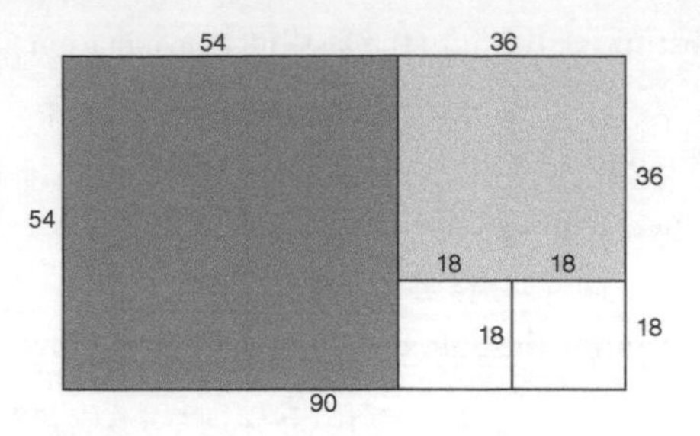

12. **gcd (90, 54) = 18.**

$$1, 2, 3, 5, 8, 13, 21, 34, 55, 89, 144, \ldots,$$

where each number is the sum of the previous two numbers: for example, 89 = 55 + 34. These numbers are usually named *Fibonacci numbers* after Leonardo Fibonacci of Pisa who mentioned them in a problem in the year 1202, although their origins are older than this. For example, using Euclid's algorithm to find gcd (89, 55) requires nine steps (see Figure 13):

$$\begin{aligned}
\mathbf{89} &= (1 \times \mathbf{55}) + \mathbf{34} \\
\mathbf{55} &= (1 \times \mathbf{34}) + \mathbf{21} \\
\mathbf{34} &= (1 \times \mathbf{21}) + \mathbf{13} \\
\mathbf{21} &= (1 \times \mathbf{13}) + \mathbf{8} \\
\mathbf{13} &= (1 \times \mathbf{8}) \quad + \mathbf{5} \\
\mathbf{8} &= (1 \times \mathbf{5}) \quad + \mathbf{3} \\
\mathbf{5} &= (1 \times \mathbf{3}) \quad + \mathbf{2} \\
\mathbf{3} &= (1 \times \mathbf{2}) \quad + \mathbf{1} \\
\mathbf{2} &= (2 \times \mathbf{1}) \quad + \mathbf{0}.
\end{aligned}$$

Here the greatest common divisor and every quotient (except the last) are all 1, and all the non-zero remainders are themselves Fibonacci numbers.

So sometimes Euclid's algorithm works more quickly than at other times. But in spite of this, it is by far the most efficient algorithm for finding greatest common divisors in general.

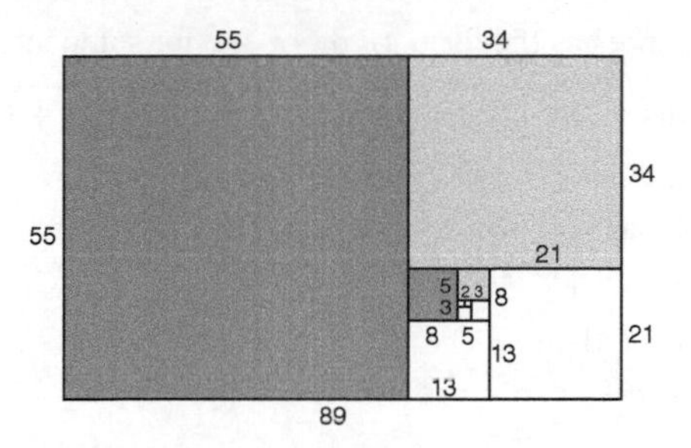

13. gcd (89, 55) = 1.

Squares

Perfect squares feature throughout number theory. As we saw in Chapter 1, the Pythagoreans have been credited with noticing that adding the first few odd numbers always gives a square: for example,

$$1 + 3 + 5 + 7 + 9 = 25 = 5^2.$$

They supposedly explained such results by drawing square diagrams similar to that in Figure 14, where the numbers of dots in the L-shaped regions are 1, 3, 5, 7, and 9. In fact, for any number k,

$$1 + 3 + 5 + \cdots + (2k - 1) = k^2.$$

Other results involving squares arise from our earlier special cases of the division rule. For example:

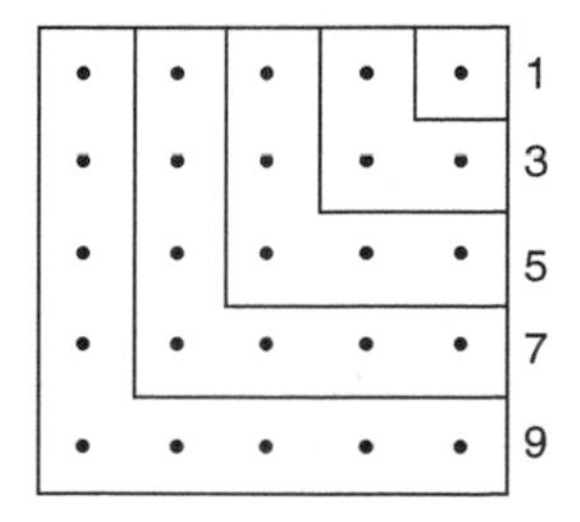

14. The sum of the first few odd numbers is a square.

Every square has the form $4n$ or $4n + 1$, for some integer n.

This is because every integer b has the form $2q$ or $2q + 1$:

if $b = 2q$, then $b^2 = 4q^2$,

which has the form $4n$ with $n = q^2$:

for example, $6^2 = 36 = 4 \times 9$;

if $b = 2q + 1$, then

$$b^2 = (2q+1)^2 = 4q^2 + 4q + 1 = 4(q^2 + q) + 1,$$

which has the form $4n + 1$ with $n = q^2 + q$:

for example, $7^2 = 49 = (4 \times 12) + 1$.

So if b is even then b^2 has the form $4n$, and if b is odd then b^2 has the form $4n + 1$.

An immediate consequence of this result is that none of the numbers

$$11, 111, 1111, 11111, \ldots$$

can be a square. This is because these numbers all have the form $4n + 3$, for some integer n: for example, $1111 = (4 \times 277) + 3$ and $11111 = (4 \times 2777) + 3$.

But we can say a little more. We've just seen that if $b = 2q + 1$, then

$$b^2 = (2q+1)^2 = 4q^2 + 4q + 1 = 4q(q+1) + 1.$$

But $q(q + 1)$ is the product of two consecutive integers (one odd and the other even), and so is even. It follows that $4q(q + 1)$ is divisible by 8, and so:

The square of every odd number has the form $8n + 1$, for some integer n.

We can also demonstrate this result geometrically: see Figure 15, for the case $b = 9$. Here, b^2 dots are arranged as eight triangles with an extra dot in the centre. Now

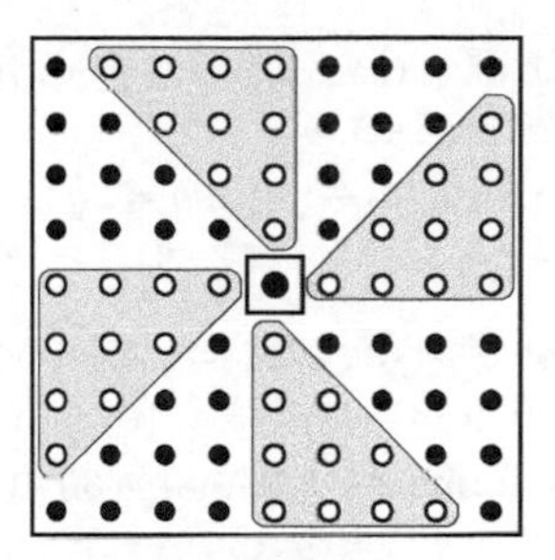

15. If b is odd, then b^2 has the form $8n+1$.

> b^2 (the total number of dots) $= 8 \times$ (the number of dots in each triangle) $+ 1$ (the central dot),

and so b^2 has the form $8n + 1$.

In Chapter 1 we saw that the squares from 1^2 to 10^2 end with 1, 4, 5, 6, 9, or 0. Is this true for all squares? By the division rule, every integer can be written as $10q + r$, where $0 \leq r \leq 9$, and so

$$(10q + r)^2 = 100q^2 + 20qr + r^2 = 10\,(10q^2 + 2qr) + r^2.$$

So the square of $10q + r$ ends with the same digit as the square of r, and so must also be 1, 4, 5, 6, 9 or 0. It follows that no perfect square can end in 2, 3, 7, or 8.

We can also use the division rule to obtain results involving cubes. A single example will give the idea.

> Every cube has the form $9n, 9n + 1,$ or $9n + 8$.

This is because every integer b has the form $3q, 3q + 1,$ or $3q + 2$:

> if $b = 3q$, then $b^3 = 27q^3$,
> which has the form $9n$ with $n = 3q^3$:
> for example, $6^3 = 216 = 9 \times 24$;

$$\text{if } b = 3q + 1, \text{ then } b^3 = (3q+1)^3 = 27q^3 + 27q^2 + 9q + 1$$
$$= 9(3q^3 + 3q^2 + q) + 1,$$
which has the form $9n + 1$ with $n = 3q^3 + 3q^2 + q$:
for example, $4^3 = 64 = (9 \times 7) + 1$;

$$\text{if } b = 3q + 2, \text{ then } b^3 = (3q+2)^3 = 27q^3 + 54q^2 + 36q + 8$$
$$= 9(3q^3 + 6q^2 + 4q) + 8,$$
which has the form $9n + 8$ with $n = 3q^3 + 6q^2 + 4q$:
for example, $5^3 = 125 = (9 \times 13) + 8$.

So if $b = 3q$ then b^3 has the form $9n$, if $b = 3q + 1$ then b^3 has the form $9n + 1$, and if $b = 3q + 2$ then b^3 has the form $9n + 8$.

We conclude this section by stating without proof an intriguing result that links squares and cubes. In Chapter 1 we asked whether the sum of the first few positive cubes must always be a perfect square. But it can be proved that, for any number n,

$$1^3 + 2^3 + 3^3 + \cdots + n^3 = (1 + 2 + 3 + \cdots + n)^2,$$

and so the result is indeed true: for example,

$$1^3 + 2^3 + 3^3 + \cdots + 10^3 = 1 + 8 + 27 + 64 + 125 + 216$$
$$+343 + 512 + 729 + 1000 = 3025$$
$$\text{and } (1 + 2 + 3 + \ldots + 10)^2 = 55^2 = 3025.$$

Divisor tests

In our decimal counting system we can write any whole number, such as 47,972, as a sum of powers of 10 (where 10^0 is taken to be 1):

$$47,972 = 40,000 + 7000 + 900 + 70 + 2$$
$$= (4 \times 10^4) + (7 \times 10^3) + (9 \times 10^2) + (7 \times 10^1) + (2 \times 10^0),$$

and in general we can write any positive number

$$n = a_k\, a_{k-1} \ldots a_2\, a_1\, a_0$$

in the form

$$n = (a_k \times 10^k) + (a_{k-1} \times 10^{k-1}) + \cdots + (a_2 \times 10^2) + (a_1 \times 10^1) + (a_0 \times 10^0).$$

Because our decimal system is a place-value system, we need to use only the ten digits

1, 2, 3, 4, 5, 6, 7, 8, 9, and 0,

with the two 7s in 47,972 standing for 7000 and 70. We can therefore carry out our calculations in columns, with the columns representing units, tens, hundreds, thousands, . . . , as we move from right to left.

In a similar way, we can write any number as a sum of powers of 2 (the binary counting system used in computing), or of 12 (the duodecimal system used for feet and inches, and formerly in Britain for shillings and pence), or of any other integer greater than 1, and many statements about the decimal system have their analogues in these other systems too. For simplicity we'll consider only decimal numbers here.

In this section we'll see some tests that tell us very quickly whether a given positive number is divisible by the integers 2, 3, 4, 5, 8, 9, 10, 11, and 25, and we'll also explain why they work. We'll also show how to test for divisibility by some larger numbers.

As we saw in Chapter 1, we can easily test the number $n = a_k a_{k-1} \ldots a_2 a_1 a_0$ for divisibility by 10 and by 5 by checking its last digit, a_0:

Divisibility by 10: n is divisible by 10 if and only if its last digit is 0.
Divisibility by 5: n is divisible by 5 if and only if its last digit is 0 or 5.

For example, 19,720 is divisible by both 10 and 5, and 19,725 is divisible by 5.

These are because if

$$n = (a_k \times 10^k) + (a_{k-1} \times 10^{k-1}) + \ldots + (a_2 \times 10^2)$$
$$+ (a_1 \times 10^1) + (a_0 \times 10^0),$$

then 10 divides all the powers of 10, except for 10^0 (= 1), and so n is divisible by 10 if and only if 10 divides a_0—that is, $a_0 = 0$, and so n ends in 0. Likewise, 5 also divides all powers of 10, except for 10^0, and so n is divisible by 5 if and only if 5 divides a_0—that is, $a_0 = 0$ or 5, and so n ends in 0 or 5.

We can likewise test for divisibility by 25 by checking its last *two* digits, a_1 and a_0:

> *Divisibility by* 25: n is divisible by 25 if and only if its last two digits are 00, 25, 50, or 75.

For example, 19,725 is divisible by 25.

This is because 25 divides all the powers of 10, except for 10^1 and 10^0, and so n is divisible by 25 if and only if 25 divides the two-digit number a_1a_0 (= $10a_1 + a_0$)—that is, n ends in 00, 25, 50, or 75.

In Chapter 1 we saw that.

> *Divisibility by* 2: n is divisible by 2 if and only if its last digit is 2, 4, 6, 8, or 0.

For example, 19,726 is divisible by 2.

This is because 2 divides all the powers of 10, except for 10^0, and so n is divisible by 2 if and only if 2 divides a_0—that is, its last digit is even.

We can likewise test for divisibility by 4 or by 8 by checking the last *two* or *three* digits:

Divisibility by 4: n is divisible by 4 if and only if its last two digits are 00, 04, 08, . . . , 92, or 96.

Divisibility by 8: n is divisible by 8 if and only if its last three digits are 000, 008, 016, . . . , 984, or 992.

For example, 19,724 is divisible by 4 and 19,728 is divisible by 8.

These are because 4 divides all the powers of 10, except for 10^1 and 10^0, and so n is divisible by 4 if and only if 4 divides the two-digit number a_1a_0. Likewise, 8 divides all the powers of 10, except for 10^2, 10^1, and 10^0, and so n is divisible by 8 if and only if 8 divides the three-digit number $a_2a_1a_0$.

We next turn our attention to divisibility by 3 and by 9.

Divisibility by 3: n is divisible by 3 if and only if the sum of its digits is divisible by 3.

Divisibility by 9: n is divisible by 9 if and only if the sum of its digits is divisible by 9.

For example, 19,725 is divisible by 3 because the sum of its digits is $1 + 9 + 7 + 2 + 5 = 24$, which is divisible by 3, and 19,728 is divisible by 9 because the sum of its digits is 27, which is divisible by 9.

These are because 3 and 9 divide all of the numbers 9, 99, 999, . . . , and so each power of 10 leaves a remainder of 1 when divided by 3 or 9. It follows that when

$$n = (a_k \times 10^k) + (a_{k-1} \times 10^{k-1}) + \cdots + (a_2 \times 10^2) + (a_1 \times 10^1) + (a_0 \times 10^0)$$

is divided by 3 or 9, the resulting remainder is simply the sum of its digits,

$$a_k + a_{k-1} + a_{k-2} + \cdots + a_0,$$

and so n is divisible by 3 or 9 if and only if this 'digital sum' is divisible by 3 or 9.

We can use a similar idea to test for divisibility by 11:

> *Divisibility by* 11: n is divisible by 11 if and only if the alternating sum of its digits is divisible by 11.

Here the 'alternating sum' is $a_k - a_{k-1} + a_{k-2} - \cdots \pm a_0$. For example, 19,723 is divisible by 11 because its alternating sum $1 - 9 + 7 - 2 + 3 = 0$ is divisible by 11.

The method works because the powers of 10 leave remainders of 1 and -1 alternately when divided by 11, so n is divisible by 11 if and only if this alternating sum is divisible by 11.

We can test for divisibility by other numbers by combining these results. For example, we can test for divisibility by 6 by testing whether n is divisible by 2 and also by 3, and we can test for divisibility by 88 by testing whether n is divisible by 8 and also by 11. In general, if n is divisible by both a and b, where a and b are coprime, then n is divisible by $a \times b$.

Casting out nines

We'll conclude this chapter with an ancient method for checking the accuracy of an arithmetical calculation. Known as 'Casting out nines', it is based on the fact that a number and its digital sum leave the same remainder when divided by 9. The method seems to have developed in India around the year 1000, and was later transmitted by Islamic scholars to Europe where versions of it are still sometimes used—for example, in bookkeeping. It's similar in idea to the final 'check digit' of a book's 13-digit ISBN number, which is included to provide a check on the accuracy of the first twelve digits.

Consider a number such as 4567. Its remainder after division by 9 is the same as that of its digital sum $4 + 5 + 6 + 7 = 22$, which in turn is the same as that of *its* digital sum, $2 + 2 = 4$. Similarly, the remainder when the number 6537 is divided by 9 is the same as that of its digital sum $6 + 5 + 3 + 7 = 21$, which in turn is the same as that of *its* digital sum, $2 + 1 = 3$. And in general, we can similarly reduce any given number n to a single-digit number, called its 'digital root', which is the remainder when we divide n by 9. (If the digital root is 9, we replace it by 0.)

In most cases, we can use these digital roots to verify the correctness (or otherwise) of an arithmetical calculation. To illustrate the idea we'll start with the *incorrect* addition sum

$$4567 + 6537 = 11{,}144.$$

Here the digital root of 4567 is 4 and of 6537 is 3, so the remainder on dividing the left-hand side by 9 is 4 + 3, which is 7. But the digital root of 11,144 is 2, so the remainder on dividing the right-hand side by 9 is 2. Because these remainders disagree, the calculation must be incorrect—the correct answer is 11,104. And this happens in general: whenever the digital roots of the two sides differ, then we know (without carrying out the calculation) that the answer is wrong.

However, the method sometimes fails: for example, consider the addition sum

$$4567 + 6537 = 11{,}194.$$

Here, the digital root of each side of the equation is 7, so the two digital roots of the two sides agree, even though the calculation is incorrect. In these cases, the correct and incorrect answers must always differ by a multiple of 9.

This method of casting out nines can be used as a check whenever we wish to add, subtract, or multiply whole numbers: in each case, we replace each number by its digital root, and check the same calculation on these digital roots. As an example of multiplication, consider the product

$$23 \times 89 = 2037.$$

The digital roots of the numbers on the left-hand side are 5 and 8, and the digital root of their product $5 \times 8 = 40$ is 4. But the digital root of the right-hand side is 3, so the digital roots disagree and the calculation is incorrect—the correct answer is 2047.

In medieval times, calculators would draw the diagram in Figure 16, where the numbers on the left and right are the digital roots 5 and 8, the number at the top is the digital root of the given answer, 3, and the number at the bottom is the digital root 4, obtained by multiplying the digital roots 5 and 8. When the numbers at the top and bottom disagree, as here, the calculation is incorrect. When they agree, the calculation is usually correct.

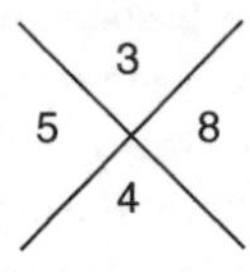

16. Casting out nines.

(a)

(b)

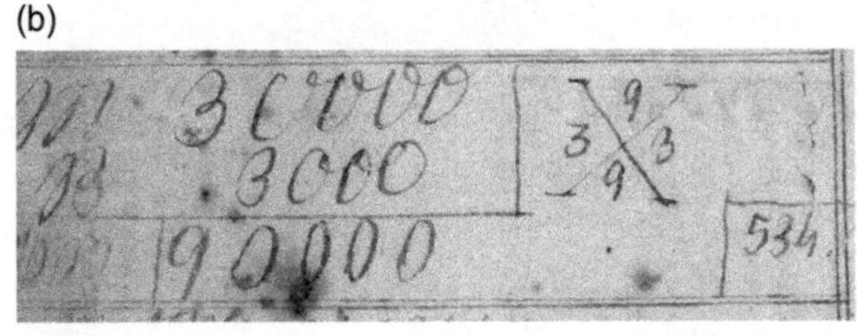

17. A German postage stamp commemorates Adam Riese. An example from Abraham Lincoln's 'Cyphering book'.

This large cross was used for hundreds of years: Figure 17 shows its use by the 16th-century German Rechenmeister (master of 'reckoning') Adam Riese, and in an arithmetical calculation by Abraham Lincoln. Eventually the cross decreased in size and became the multiplication sign that we use today.

Chapter 3
Prime-time mathematics

As we saw in Chapter 1, prime numbers lie at the heart of number theory, and we'll now explore some of their properties. We'll see how to generate prime numbers and show that the list of primes is never-ending. We'll see that there's essentially just one way of splitting any given number into its prime factors, and we'll introduce some important types of primes. Our explorations will spill over into Chapter 7 where we'll describe some more general issues, such as how the prime numbers are distributed and whether we can find progressions of equally spaced prime numbers.

We recall from Chapter 1 that a *prime number* is a number, greater than 1, whose only factors (divisors) are itself and 1, and a number that's not prime is *composite*. A list of all the prime numbers up to 1000 appears in Table 1. Notice that some primes appear to cluster quite closely together, such as 101, 103, 107, and 109, whereas others, such as 113 and 127, are more widely spaced.

The sieve of Eratosthenes

How might we generate such a list of prime numbers? An early method was to use the *sieve of Eratosthenes*, named after Eratosthenes of Cyrene who lived in the 3rd century BC and is also celebrated for estimating the circumference of the Earth. His idea

Table 1. The prime numbers up to 1000

2	3	5	7	11	13	17	19	23	29	31	37
41	43	47	53	59	61	67	71	73	79	83	89
97	101	103	107	109	113	127	131	137	139	149	151
157	163	167	173	179	181	191	193	197	199	211	223
227	229	233	239	241	251	257	263	269	271	277	281
283	293	307	311	313	317	331	337	347	349	353	359
367	373	379	383	389	397	401	409	419	421	431	433
439	443	449	457	461	463	467	479	487	491	499	503
509	521	523	541	547	557	563	569	571	577	587	593
599	601	607	613	617	619	631	641	643	647	653	659
661	673	677	683	691	701	709	719	727	733	739	743
751	757	761	769	773	787	797	809	811	821	823	827
829	839	853	857	859	863	877	881	883	887	907	911
919	929	937	941	947	953	967	971	977	983	991	997

was to imagine putting the positive integers into a sieve and letting all the composite numbers fall though the holes, leaving just the primes. For example, to produce the above table of primes up to 1000 we list all the numbers from 2 to 1000 and then systematically sift out the multiples of 2, then the multiples of 3, and then those of 5, 7, 11, . . . for each successive prime number.

To see how this sifting process works in practice, we'll find all the primes up to 100. We start by listing all the numbers up to 100 and cross out 1. We then leave in 2, but cross out all its larger multiples—that is, 4, 6, 8, . . . , 98, 100. This gives us the following table:

×	2	3	×	5	×	7	×	9	×
11	×	13	×	15	×	17	×	19	×
21	×	23	×	25	×	27	×	29	×
31	×	33	×	35	×	37	×	39	×
41	×	43	×	45	×	47	×	49	×
51	×	53	×	55	×	57	×	59	×
61	×	63	×	65	×	67	×	69	×
71	×	73	×	75	×	77	×	79	×
81	×	83	×	85	×	87	×	89	×
91	×	93	×	95	×	97	×	99	×

We then leave in the next number, which is 3, but cross out those of its larger multiples that remain—that is, 9, 15, 21, . . . , 93, 99. Note that the other multiples of 3, such as 6, 12, and 18, were already crossed out at the previous stage.

×	**2**	**3**	×	5	×	7	×	×	×
11	×	13	×	×	×	17	×	19	×
×	×	23	×	25	×	×	×	29	×
31	×	×	×	35	×	37	×	×	×
41	×	43	×	×	×	47	×	49	×
×	×	53	×	55	×	×	×	59	×
61	×	×	×	65	×	67	×	×	×
71	×	73	×	×	×	77	×	79	×
×	×	83	×	85	×	×	×	89	×
91	×	×	×	95	×	97	×	×	×

Repeating the process for the larger multiples of 5, and then 7, leaves:

×	**2**	**3**	×	**5**	×	**7**	×	×	×
11	×	13	×	×	×	17	×	19	×
×	×	23	×	×	×	×	×	29	×
31	×	×	×	×	×	37	×	×	×
41	×	43	×	×	×	47	×	×	×
×	×	53	×	×	×	×	×	59	×
61	×	×	×	×	×	67	×	×	×
71	×	73	×	×	×	×	×	79	×
×	×	83	×	×	×	×	×	89	×
×	×	×	×	×	×	97	×	×	×

At this stage we might worry that this sifting process could take a long time to complete, but this isn't the case: every multiple of 11 (other than 11 itself) has already been crossed out, as has every multiple of 13 and of all larger primes, so we already have our full list of primes up to 100. We don't need to check the multiples of any primes above 10, because if a number from 2 to 100 has a prime factor p that's greater than 10, then it must also have a prime factor q that's less than 10; this is because if q were also greater than 10, then $p \times q$ would be greater than 100.

In general, to find all the primes up to a given number n we need to cross out only the larger multiples of the primes up to its square root, $\sqrt{n}$. For example, to find all the primes up to 200 we cross out only the larger multiples of the primes up to $\sqrt{200} = 14.142\ldots$ (these are the primes 2, 3, 5, 7, 11, and 13), and to find all the primes up to 1000 we cross out only the larger multiples of the primes up to $\sqrt{1000}$ (these are the primes up to 31).

Primes go on for ever

The list of prime numbers continues for ever: there is no largest prime. As we mentioned in Chapter 1, this was proved in the 3rd century BC by Euclid in Book IX of his *Elements*:

> There are infinitely many primes.

At first sight it may seem difficult to see how we might attempt to prove such a result, because if we're given a specific prime number, such as 997, it's not immediately obvious what the next prime would be. (It's actually 1009.)

Euclid's proof of his theorem is considered one of the great classics of mathematics, and is a proof by contradiction (sometimes called *reductio ad absurdum*). This means that we suppose the opposite result—that there are only *finitely many* primes—and then show how that supposition leads us to a contradictory statement. His method was based on the fact that, given any collection of primes, we can always find a new one and so, by continually repeating this process with the new list, we can carry on generating new primes for ever. This contradiction establishes the result.

To illustrate the idea behind this process we'll take a collection of prime numbers, multiply them together, add 1, and then look at the resulting number. For example, if we start with 2, 3, and 5, we have

$$(2 \times 3 \times 5) + 1 = 31,$$

which is a prime number that wasn't in our original list. Or if we start with 2, 5, and 11, we have

$$(2 \times 5 \times 11) + 1 = 111,$$

which isn't a prime number but which splits as 3×37, giving us two new primes: 3 and 37. In each case we've generated at least one new prime number that wasn't in our original list.

A more formal proof is as follows. We'll suppose that there are *finitely many* primes, which we'll call $p_1, p_2, \ldots,$ and p_n, and we'll form the number

$$N = (p_1 \times p_2 \times \cdots \times p_n) + 1.$$

Now, each of these primes divides their product, $p_1 \times p_2 \times \cdots \times p_n$, and so cannot also divide N because this would leave a remainder of 1. So either N is a new prime, or N is a composite number in which case it must split into new primes. In either case, there must exist a prime that's different from $p_1, p_2, \ldots,$ and p_n, and this contradicts our assumption that these were the only primes. Our original assumption must therefore be false, and so there are infinitely many primes.

Factorizing into primes

Earlier I mentioned that the prime numbers are the building blocks for our counting system, and we'll now explore this idea further.

Starting from any collection of primes, we can build up further numbers by multiplication: for example, beginning with the primes 2 and 3 we can construct the numbers

$$2 \times 3 = 6, \quad 2 \times 2 \times 2 \times 2 \times 2 = 32, \quad 3 \times 3 \times 3 \times 3 = 81,$$

$$2 \times 2 \times 3 \times 3 \times 3 = 108,$$

and any other number of the form $2^k \times 3^l$, where $k \geq 0$ and $l \geq 0$.

Turning things around, we see that every integer greater than 1 splits into primes, because if the number n is composite then we can write $n = a \times b$ for smaller integers a and b: for example, $50 = 10 \times 5$. If a is composite, then we can split it up further, and similarly for b. Continuing in this way, after a finitely many steps we'll eventually obtain n as a product of primes.

Moreover, if we're given a number such as 108, then we can split it up in more than one way: for example,

$$108 = 4 \times 27 = 2 \times 2 \times 3 \times 3 \times 3$$
$$108 = 9 \times 12 = 3 \times 3 \times 2 \times 2 \times 3.$$

In the same way, we can write

$$630 = 63 \times 10 = (9 \times 7) \times (2 \times 5) = 3 \times 3 \times 7 \times 2 \times 5$$
$$630 = 6 \times 105 = (2 \times 3) \times (7 \times 15) = 2 \times 3 \times 7 \times 3 \times 5.$$

We can illustrate all these factorizations as tree diagrams (see Figure 18).

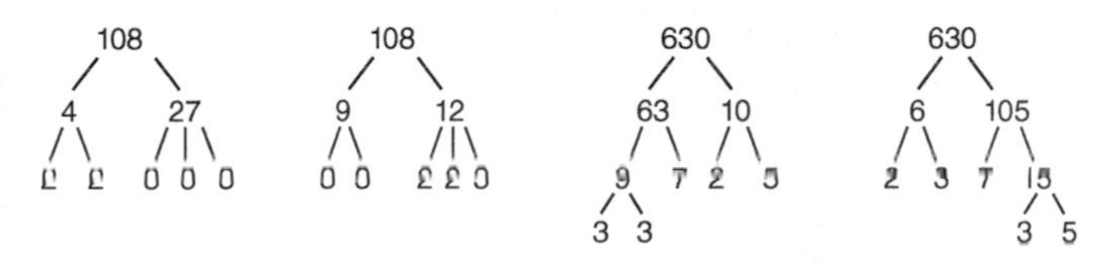

18. Factorizations of 108 and 630.

In each case the prime factors at the lower ends of the trees are exactly the same, but they appear in a different order. This suggests the following result, which is known as the 'fundamental theorem of arithmetic'. Although it must have been familiar to mathematicians for hundreds of years, it seems not to have been proved formally until Gauss did so in his *Disquisitiones Arithmeticae* of 1801.

> *Fundamental theorem of arithmetic*: Every integer greater than 1 is either a prime number or can be written as a

product of primes. Moreover, there is only one factorization into its prime factors, apart from the order in which they appear.

We can now see why we don't wish 1 to be a prime number—for, if it were, then we could write, for example,

$$6 = 2 \times 3, \quad \text{or} \quad 1 \times 2 \times 3, \quad \text{or} \quad 1 \times 1 \times 2 \times 3,$$
$$\text{or} \quad 1 \times 1 \times 1 \times 2 \times 3, \quad \text{or} \ldots,$$

so that the uniqueness of the prime factorization would be lost.

Using the fundamental theorem of arithmetic, we can now write every positive integer as a product of primes in a standard way, by listing its prime factors in increasing order of size, with each one raised to the appropriate power: for example,

$$108 = 2^2 \times 3^3, \quad 630 = 2 \times 3^2 \times 5 \times 7, \quad 1000 = 2^3 \times 5^3.$$

In general, if the prime factors of the number n are $p_1, p_2, \ldots$, and p_r, where each prime p_i is raised to the power e_i and the primes are listed in increasing order, then we write

$$n = p_1^{e_1} \times p_2^{e_2} \times \cdots \times p_r^{e_r}.$$

This is called the *canonical form* of the prime factorization of n, and we'll often write prime factorizations in this way.

To conclude this section we'll return briefly to least common multiples and greatest common divisors. In Chapter 2 we showed that lcm (90, 54) = 270 and gcd (90, 54) = 18, by listing the multiples and divisors of 90 and 54. But a quicker method is to write each number as a product of primes in canonical form:

$$90 = 2^1 \times 3^2 \times 5^1 \quad \text{and} \quad 54 = 2^1 \times 3^3 \quad \text{or} \quad = 2^1 \times 3^3 \times 5^0,$$

where the extra factor 5^0 (which equals 1) is introduced so that the primes that appear (2, 3, and 5) are the same. Then:

to find lcm (90, 54), we look at each prime factor in turn, take the *larger* power that appears—that is, 2^1, 3^3, and

5^1—and multiply them to give $2^1 \times 3^3 \times 5^1 = 270$;

to find gcd (90, 54), we look at each prime factor in turn, take the *smaller* power that appears—that is, 2^1, 3^2, and 5^0—and multiply them to give $2^1 \times 3^2 \times 5^0 = 18$.

In general, if a and b are written in canonical form as

$$a = p_1^{e_1} \times p_2^{e_2} \times \cdots \times p_r^{e_r} \quad \text{and} \quad b = p_1^{f_1} \times p_2^{f_2} \times \cdots \times p_r^{f_r}$$

(where we introduce the exponent 0 whenever a particular prime doesn't appear), then

$$\operatorname{lcm}(a, b) = p_1^{\max(e_1, f_1)} \times p_2^{\max(e_2, f_2)} \times \cdots \times p_r^{\max(e_r, f_r)},$$

where max (e, f) is the larger of e and f, and

$$\gcd(a, b) = p_1^{\min(e_1, f_1)} \times p_2^{\min(e_2, f_2)} \times \cdots \times p_r^{\min(e_r, f_r)},$$

where min (e, f) is the smaller of e and f.

We can now revisit a result from Chapter 2—that, for any numbers a and b,

$$\operatorname{lcm}(a, b) \times \gcd(a, b) = a \times b.$$

To explain why this is true we note first that, for any numbers e and f,

$$\max(e, f) + \min(e, f) = e + f$$

—for example, $\max(2, 3) + \min(2, 3) = 3 + 2$. Then the power of the first prime factor p_1 on the left-hand side of the result that we're trying to prove is

$$p_1^{\max(e_1, f_1)} \times p_1^{\min(e_1, f_1)} = p_1^{\max(e_1, f_1) + \min(e_1, f_1)} = p_1^{e_1 + f_1} = p_1^{e_1} \times p_1^{f_1},$$

which is the power of p_1 on the right-hand side. So the powers of p_1 on the two sides match, and the same is true for all the other prime factors. So the two sides are equal.

Searching for primes

Are there any formulas for producing primes? Here are some historical attempts at finding them.

Euler's primes

Leonhard Euler was obsessed with prime numbers, as we'll see throughout this book. In 1771 he investigated the formula

$$41 - n + n^2,$$

and made the surprising observation that if we substitute every number n from 1 to 40 in turn, then we always get primes:

$$n = 1 \text{ gives } 41 - 1 + 1^2 = 41,$$
$$n = 2 \text{ gives } 41 - 2 + 2^2 = 43,$$
$$n = 3 \text{ gives } 41 - 3 + 3^2 = 47,$$

and so on, up to

$$n = 39 \text{ gives } 41 - 39 + 39^2 = 1523,$$
$$n = 40 \text{ gives } 41 - 40 + 40^2 = 1601.$$

These are all prime numbers, but the formula may also fail to produce primes: for example,

$$n = 41 \text{ gives } 41 - 41 + 41^2 = 1681 = 41 \times 41,$$
$$n = 42 \text{ gives } 41 - 42 + 42^2 = 1763 = 41 \times 43.$$

Likewise, the formula

$$1601 - 79n + n^2$$

produces prime numbers for every number n from 1 to 79, but not for 80 or many larger numbers. So neither of these formulas always gives primes.

Mersenne primes

Particularly important among the prime numbers, for reasons that will soon become clear, are those that are one less than a power of 2, such as

$$2^2 - 1 = 3, \quad 2^3 - 1 = 7, \quad 2^5 - 1 = 31, \quad \text{and} \quad 2^7 - 1 = 127.$$

Numbers of the form $2^n - 1$ are called *Mersenne numbers* and, as we saw in Chapter 1, not all of them are prime. To explore which ones have this property, let's draw up a table of the first few Mersenne numbers.

Table 2. Numbers of the form $2^n - 1$

n	1	2	3	4	5	6	7	8
$2^n - 1$	1	3	7	15	31	63	127	255
n	9	10	11	12	13	14	15	16
$2^n - 1$	511	1023	2047	4095	8191	16,383	32,767	65,535

It may seem tempting, on looking at these numbers, to speculate that $2^n - 1$ is a prime number when n is prime, and that it's composite when n is composite. It's certainly true that

$2^n - 1$ is composite when n is composite,

because if r divides n (where $r \geq 2$), then it can be shown that $2^r - 1$ must divide $2^n - 1$, and so $2^n - 1$ is composite: for example, when $n = 12$ and $r = 3$, we have

$$2^{12} - 1 = (2^3 - 1) \times (2^9 + 2^6 + 2^3 + 1).$$

But it isn't always true that $2^n - 1$ is prime when n is prime: for example,

$$2^{11} - 1 = 2047 = 23 \times 89$$
$$2^{23} - 1 = 8,388,607 = 47 \times 178,481.$$

Mersenne studied these numbers in 1644, and claimed that $2^n - 1$ is prime for each of the values

$$n = 2, 3, 5, 7, 13, 17, 19, 31, 67, 127, \text{and } 257.$$

The fact that $2^{31} - 1 = 2,147,483,647$ is prime was first confirmed by Euler in 1750, and in 1876 the French mathematician Edouard Lucas showed that the following 39-digit Mersenne number is also prime:

$$2^{127} - 1$$
$$= 170{,}141{,}183{,}460{,}469{,}231{,}731{,}687{,}303{,}715{,}884{,}105{,}727.$$

So Mersenne was correct when $n = 31$ and $n = 127$. But he was wrong in the cases $n = 67$ and $n = 257$, both of which give composite numbers: for example,

$$2^{67} - 1 = 147{,}573{,}952{,}589{,}676{,}412{,}927$$
$$= 193{,}707{,}721 \times 761{,}838{,}257{,}287.$$

He omitted the cases $n = 61$, $n = 89$, and $n = 107$, which also give rise to primes, but when we consider the large numbers involved, he may surely be forgiven. A method for testing whether a given Mersenne number is prime is given in the next chapter.

There's an amusing story about the number $2^{67} - 1$. For many years mathematicians had struggled to determine whether this number is prime, as Mersenne had claimed, and in 1875 Lucas showed that it was composite but couldn't find its factors. It was Frank Nelson Cole, a professor at Columbia University in New York, who found them after three years of systematic searching on Sunday afternoons. He presented his results on 31 October 1903 at a meeting of the American Mathematical Society, when he quietly walked into the lecture room and in complete silence painstakingly calculated 2^{67} and subtracted 1 on one side of the blackboard. He then walked over to the other side of the board and, again in complete silence, calmly multiplied the numbers 193,707,721 and 761,838,257,287, obtaining the same answer, and quietly returned to his seat. The audience erupted and gave him a standing ovation and vigorous applause. At no stage had he uttered a single word!

Why are Mersenne primes important? In the continuing search for new and larger prime numbers it's been usual to turn to Mersenne numbers, because all recently discovered large primes have been of this form—there's even been a postage stamp celebrating the discovery of one of them (see Figure 19). Indeed, in recent years a new international collaborative online

19. A postage stamp celebrates the discovery in 2001 of the 39th Mersenne prime.

project has developed to track down such numbers. Involving thousands of mathematicians and computer enthusiasts, it's called GIMPS (the Great Internet Mersenne Prime Search), and since it started around 1996 it has found seventeen of them. It's not known whether there are infinitely many Mersenne primes.

At the time of writing, fifty-one Mersenne primes have been found. The most recent of these, the largest prime number currently known, was found in December 2018. It is $2^{82,589,933} - 1$, has almost 25 million digits, and to print it out in full at 75 digits per line and 50 lines per page would take many thousands of pages!

Perfect numbers

Another important feature of Mersenne primes is their connection with perfect numbers. We recall from Chapter 1 that a number N is *perfect* if N is the sum of its proper factors (those that are less than N). For example, you've seen that

> 6 is perfect, because its proper factors (1, 2, and 3) add up to 6, and 28 is perfect, because its proper factors (1, 2, 4, 7, and 14) add up to 28,

and that the next two perfect numbers are 496 and 8128. There are then no more of them until 33,550,336.

Is there a formula for perfect numbers? To find out, let's factorize the ones we already know:

$$\begin{aligned}
6 &= 2 \times 3 = 2^1 \times (2^2 - 1), \\
28 &= 4 \times 7 = 2^2 \times (2^3 - 1), \\
496 &= 16 \times 31 = 2^4 \times (2^5 - 1), \\
8128 &= 64 \times 127 = 2^6 \times (2^7 - 1), \\
33,550,366 &= 4096 \times 8191 = 2^{12} \times (2^{13} - 1).
\end{aligned}$$

These numbers all have the form $2^{n-1} \times (2^n - 1)$, where the number in parentheses is a Mersenne prime.

In Book IX of his *Elements*, Euclid investigated perfect numbers and found that the number

$$N = 2^{n-1} \times (2^n - 1)$$

is indeed perfect whenever $2^n - 1$ is a prime. To see why, let $p = 2^n - 1$. Then the proper factors of $N = 2^{n-1} \times p$ are

$1, 2, 2^2, 2^3, \ldots,$ and 2^{n-1} (those that don't involve p)

$p, 2p, 2^2p, 2^3p, \ldots,$ and $2^{n-2}p$ (those that do involve p).

What's the sum of all these factors? Using the fact that

$1 + 2 + 2^2 + \cdots + 2^r = 2^{r+1} - 1$, for any number r,

we can write

$$1 + 2 + 2^2 + \cdots + 2^{n-1} = 2^n - 1$$

$$1 + 2 + 2^2 + \cdots + 2^{n-2} = 2^{n-1} - 1.$$

So we find, after some calculation, that the sum of all the proper factors is

$$\begin{aligned}
&(1 + 2 + 2^2 + \cdots + 2^{n-1}) + (1 + 2 + 2^2 + \cdots + 2^{n-2})p \\
&\quad = (2^n - 1) + (2^{n-1} - 1)p \\
&\quad = (2^n - 1) + (2^{n-1} - 1)(2^n - 1) \\
&\quad = \{1 + (2^{n-1} - 1)\} \times (2^n - 1) = 2^{n-1} \times (2^n - 1),
\end{aligned}$$

which equals N.

So $N = 2^{n-1} \times (2^n - 1)$ is perfect.

Are *all* perfect numbers of this type, or might there be some others? In 1638 the French philosopher René Descartes wrote to Mersenne saying that he believed that all *even* perfect numbers must indeed be of Euclid's form, $2^{n-1} \times (2^n - 1)$, for some number n, and this was subsequently proved by Euler and published posthumously. Euler also proved that every even perfect number must end with either 6 or 8: for example, the next three perfect numbers are

8,589,869,056, 137,438,691,328, 2,305,843,008,139,952,128.

But what about odd perfect numbers? Descartes conjectured that these cannot exist, and it remains unknown to this day whether he was correct. But if there were any odd perfect numbers, then they would certainly be difficult to find. For a start, it's known that any odd perfect number must be at least 10^{1500}, that it must have at least 101 prime factors with at least ten different ones, and that it must have the form $12n + 1$, $468n + 117$, or $324n + 81$, for some integer n. Few mathematicians consider their existence likely.

Fermat primes

Having just looked at prime numbers of the form $2^n - 1$, let's now turn our attention to numbers of the form $2^n + 1$. As we saw in Chapter 1, Pierre de Fermat investigated these numbers (now called *Fermat numbers*) and found five of them that are prime:

$$2^1 + 1 = 3, \quad 2^2 + 1 = 5, \quad 2^4 + 1 = 17, \quad 2^8 + 1 = 257,$$
$$2^{16} + 1 = 65{,}537.$$

So is it true that $2^n + 1$ is prime when, and only when, n is a power of 2? Let's draw up another table:

Table 3. Numbers of the form $2^n + 1$

n	**1**	**2**	3	**4**	5	6	7	**8**
$2^n + 1$	**3**	**5**	9	**17**	33	65	129	**257**
n	9	10	11	12	13	14	15	**16**
$2^n + 1$	513	1025	2049	4097	8193	16,385	32,769	**65,537**

It certainly seems as though this is the case.

We first show that if $2^n + 1$ is prime, then n must be a power of 2. To do this, we note that if n had an odd factor (other than 1), then $2^n + 1$ would be composite. For, if we write $n = mr$, where m is odd, then it can be shown that $2^r + 1$ is a divisor of $2^n + 1$: for example, if $n = 12$, $m = 3$, and $r = 4$, then

$$2^{12} + 1 = (2^4 + 1) \times (2^8 - 2^4 + 1).$$

It follows that if $2^n + 1$ is prime, then n can have no odd factor, and so must be a power of 2.

Although Fermat was convinced that if n is a power of 2, then $2^n + 1$ must be prime, he confessed that he was unable to prove this, or even to show that the next one, the ten-digit number

$$2^{32} + 1 = 4,294,967,297,$$

is prime. In desperation, he wrote around to fellow mathematicians challenging them to prove his 'distinguished theorem'. At the time, no-one could do so.

Euler learned of Fermat's challenge in 1729, and after much tedious experimentation he showed it to be divisible by 641—in fact,

$$2^{32} + 1 = 641 \times 6,700,417.$$

He later discovered that if $2^{32} + 1$ had any prime factors, then they would have to be of the form $64k + 1$, for some integer k. This limits the search considerably, because the smallest prime numbers of this type are 193, 257, 449, and 557 (none of which divides $2^{32} + 1$), and 641 (which does, as we've seen).

Is the next Fermat number prime? This number is

$$2^{64} + 1 = 18{,}446{,}744{,}073{,}709{,}551{,}617.$$

Euler showed that any prime factors of $2^{64} + 1$ must be of the form $128k + 1$, for some integer k. Unfortunately, he could find no such prime number that divides $2^{64} + 1$, and he eventually gave up his search. But it turns out that $2^{64} + 1$ is indeed composite, with 274,177 as its smallest prime factor, so it's not surprising that Euler missed it. In fact,

$$2^{64} + 1 = 274{,}177 \times 67{,}280{,}421{,}310{,}721.$$

Worse was to come: the next twenty-six Fermat numbers were also found to be composite. Indeed, no-one has ever found another Fermat number that's prime, and so Fermat's conjecture has turned out to be a rather unfortunate one.

A geometrical digression

Fermat primes turn up in unexpected places. A celebrated example from geometry arises in the construction of *regular polygons*—those polygons whose sides all have the same length and whose angles are all the same, such as an equilateral triangle, a square, or a regular pentagon (see Figure 20).

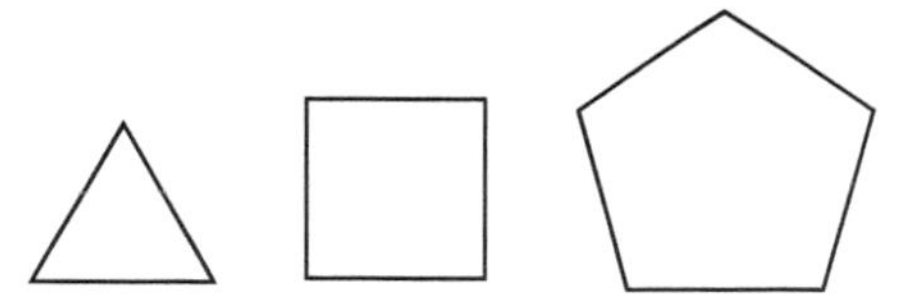

20. Some regular polygons.

The Ancient Greeks were interested in the construction of geometrical figures using only a ruler (for drawing straight lines) and compasses (for drawing circles); the ruler is assumed to be unmarked and no measuring is allowed. Indeed, the very first proposition in Euclid's *Elements*, Book I, includes the construction of an equilateral triangle. Briefly it goes as follows (see Figure 21):

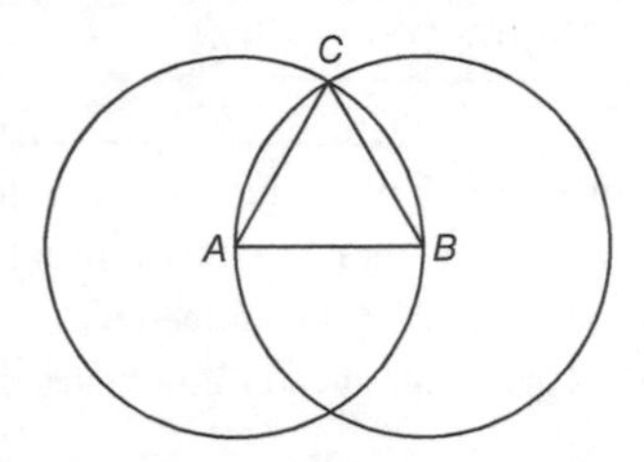

21. Constructing an equilateral triangle.

> Given a line segment AB, use the compasses to draw the circle with centre A and radius AB, and the circle with centre B and radius BA.
> These circles intersect at the point C: draw the lines AC and BC.
> Then ABC is an equilateral triangle.

Euclid also showed how to construct squares and regular pentagons, and in Book IV he combined the constructions for triangles and pentagons to produce a regular polygon with $3 \times 5 = 15$ sides. He also explained how to construct a regular polygon with $2n$ sides from one with n sides, by joining the centre of the surrounding circle to the midpoints of its sides (see Figure 22 for the case $n = 4$).

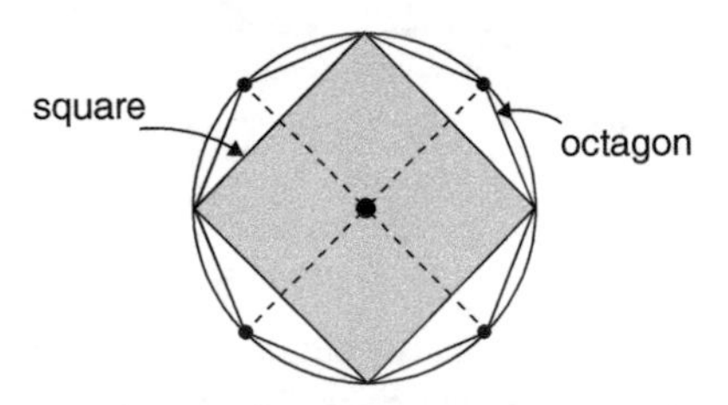

22. Doubling the number of sides of a regular polygon.

It follows that:

from an equilateral triangle (with 3 sides), we can construct regular polygons with 6, 12, 24, ... sides,

from a square (with 4 sides), we can construct regular polygons with 8, 16, 32, ... sides,

from a regular pentagon (with 5 sides), we can construct regular polygons with 10, 20, 40, ... sides.

So we can already construct regular polygons with 3, 4, 5, 6, 8, 10, 12, 15, and 16 sides.

But no-one has been able to construct regular polygons with 7, 9, 11, 13, or 14 sides, and this leads to the following question:

Which regular polygons can be constructed with an unmarked ruler and compasses?

The breakthrough was provided by the 18-year-old Carl Friedrich Gauss, who discovered a ruler-and-compasses method for constructing a regular 17-sided polygon. He then gave the following answer:

A regular polygon with n sides can be constructed if and only if n is a power of 2 multiplied by unequal Fermat primes.

It follows that one can construct regular polygons with

$$20\,(=2^2\times 5),\quad 34\,(=2\times 17),\quad \text{or}\quad 60\,(=2^2\times 3\times 5)\,\text{sides},$$

but not with

$$7,\quad 9\,(=3\times 3),\quad 11,\quad 13,\quad 14\,(=2\times 7),\quad 18\,(=2\times 3\times 3),$$
$$\text{or}\quad 100\,(=2^2\times 5\times 5)\,\text{sides}.$$

The following list gives the number of sides (up to 100) of regular polygons that can be constructed by an unmarked ruler and compasses:

3, 4, 5, 6, 8, 10, 12, 15, 16, 17, 20, 24, 30, 32, 34,
40, 48, 51, 54, 60, 64, 68, 80, 85, and 96.

It seems remarkable that Fermat primes arise in such a geometrical problem.

Two weird results

We conclude this chapter with two unexpected and rather bizarre results concerning primes.

The first of these is due to W. H. Mills who proved the following startling result in 1947:

> There is a number x, with the property that, if we raise it to the powers 3, 9, 27, 81, ... (the powers of 3) and then ignore the fractional part, we always get a prime number.

The smallest such number x, sometimes called *Mills' constant*, is approximately equal to 1.30637788: for example,

$x^3 = (1.30637788\ldots)^3 = 2.229\ldots$, giving the prime 2,
$x^9 = (1.30637788\ldots)^9 = 11.082\ldots$, giving the prime 11,
$x^{27} = (1.30637788\ldots)^{27} = 1361.000\ldots$, giving the prime 1361,
$x^{81} = (1.30637788\ldots)^{81} = 2{,}521{,}008{,}283.16\ldots$,
giving the prime 2,521,008,283.

Our second bizarre result involves the idea of a *polynomial*. This is an expression obtained by adding and subtracting multiples of powers of the variables a, b, c, etc.: for example,

$1 + 2a + 3a^3 + 4a^8$
is a polynomial in the single variable a,
$a^2 + a^3b^4 - 7ab^6 - b^9$
is a polynomial in the pair of variables a and b,
$a + 2b^4 - 3ac^2 - 4bc^2d$
is a polynomial in the four variables a, b, c, and d.

In 1976 four mathematicians (J. Jones, D. Sato, H. Wada, and D. Wiens) discovered a somewhat horrendous polynomial in twenty-six variables ($a, b, c, \ldots,$ and z), with the following remarkable properties:

> If you substitute any integers you wish for all of the twenty-six variables, and if the result you get is a positive number, then it must be a prime number.
> Moreover, *every* prime number can be obtained in this way, by making a suitable choice of the integers $a, b, c, \ldots,$ and z.

Their polynomial is

$$(k+2) \times (1 - A^2 - B^2 - C^2 - D^2 - E^2 - F^2 - G^2 - H^2 - I^2 - J^2 - K^2 - L^2 - M^2 - N^2),$$

where

$$A = wz + h + j - q, \quad B = (gk + 2g + k + 1)(h + j) + h - z,$$
$$C = 2n + p + q + z - e,$$
$$D = 16(k+1)^3 (k+2)(n+1)^2 + 1 - f^2,$$
$$E = e^3 (e+2)(a+1)^2 + 1 - o^2, \quad F = a^2y^2 + 1 - x^2 - y^2,$$
$$G = 16r^2y^4(a^2 - 1) + 1 - u^2, \quad H = a^2l^2 - l^2 - m^2 + 1,$$
$$I = \{(a + u^2(u^2 - a))^2 - 1\}(n + 4dy)^2 + 1 - (x + cu)^2,$$
$$J = ai + k - l + 1 - i, \quad K = n + l + v - y,$$
$$L = p - m + al - ln - l^2 + b(2an + 2a - n^2 - 2n - 2),$$
$$M = q + ay - py - x - y + s(2ap + 2a - p^2 - 2p - 2),$$
$$N = z + alp - lp^2 - pm + t(2ap - p^2 - 1).$$

This polynomial can take a positive value only if all the squared terms, $A^2, B^2, \ldots,$ and N^2 are 0, and so to discover prime values we need to solve fourteen simultaneous equations. Surprisingly, specific integers $a, b, c, \ldots,$ and z have never been found that give the prime number 2—but it's still possible to prove that *every* prime number can be obtained in this way!

We'll continue with our explorations of prime numbers in Chapter 7. But first, we'll investigate a different type of arithmetic.

Chapter 4
Congruences, clocks, and calendars

In 1801, in his revolutionary text *Disquisitionae Arithmeticae*, Gauss introduced the idea of 'congruence', a generalized form of equality that's sometimes popularized as 'clock arithmetic'. But its origins are much older than this, and in this chapter we'll meet the 'Chinese remainder theorem', which can be traced back to Ancient China and yet has widespread applications throughout mathematics today. Further applications of congruences include testing for Mersenne primes and finding the day of the week on which a given date falls, and the chapter ends with Gauss's celebrated law of quadratic reciprocity.

Clock arithmetic

Imagine a clock face (see Figure 23).

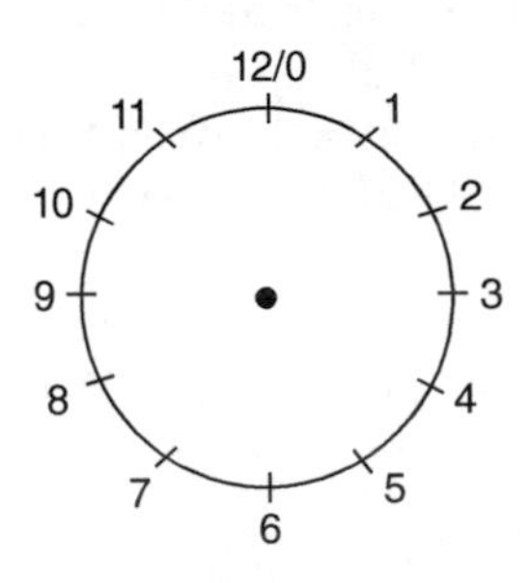

23. A 12-hour clock.

If it's now 9 o'clock, then in 6 hours it will be 3 o'clock: we'll write this as

$$9 + 6 \equiv 3 \pmod{12}.$$

Here, 'mod 12' means that we're using a 12-hour clock, so that 15 (the sum of 9 and 6) is replaced by 3, and we write '≡' instead of '=' because it's a different form of equality.

Similarly, if it's now 10 o'clock, then in 7 hours it will be 5 o'clock, and we write this as

$$10 + 7 \equiv 5 \pmod{12},$$

where 17 (the sum of 10 and 7) is replaced by 5. It turns out to be convenient to replace 12 by 0, so that if it's 8 o'clock now, then in 4 hours it will be 0 o'clock, and we write this as

$$8 + 4 \equiv 0 \pmod{12}.$$

This type of calculation is called *clock arithmetic*: if we add the hours and get a number that's 12 or larger, then we subtract 12 before giving the answer: the answer we give is then the remainder after we've divided the sum by 12, and is one of the numbers from 0 to 11.

In general, given any number n, greater than 1, we say that two integers a and b are *congruent* mod n if a and b leave the same remainder when divided by n, and we write

$$a \equiv b \pmod{n}.$$

For example, as we've seen,

$$15 \equiv 3 \pmod{12}, \quad 17 \equiv 5 \pmod{12}, \quad 12 \equiv 0 \pmod{12}.$$

The abbreviation 'mod' is short for *modulo*, and the official name for clock arithmetic is *modular arithmetic*.

It follows from the definition that $a \equiv b \pmod{n}$ whenever the difference $a - b$ is divisible by n: for example,

$$15 \equiv 9 \pmod{6}, \quad 37 \equiv 17 \pmod{10}, \quad 8 \equiv -2 \pmod{5}.$$

Since the only remainders when we divide by n can be 0, 1, 2, ..., or $n - 1$, every integer is congruent (mod n) to one of these: for example,

$$15 \equiv 3 \pmod{6}, \quad 37 \equiv 7 \pmod{10}, \quad 8 \equiv 3 \pmod{5}.$$

We can also carry out arithmetic on congruences, provided that we stick to the same modulus. For example, given the congruences $10 \equiv 3 \pmod{7}$ and $12 \equiv 5 \pmod{7}$,

we can add them to give $22 \equiv 8 \pmod{7}$, which we can rewrite as $22 \equiv 1 \pmod{7}$;

we can subtract them to give $-2 \equiv -2 \pmod{7}$, or $-2 \equiv 5 \pmod{7}$;

we can multiply them to give $120 \equiv 15 \pmod{7}$, or $120 \equiv 1 \pmod{7}$.

In general, if $a \equiv b \pmod{n}$ and $c \equiv d \pmod{n}$, then

$$a + c \equiv b + d \pmod{n}, \; a - c \equiv b - d \pmod{n}, \; ac \equiv bd \pmod{n}.$$

We can also add, subtract, or multiply a congruence by a constant integer: for example, starting with $10 \equiv 3 \pmod{7}$ we can add 2, subtract 2, or multiply by 2, to give

$$12 \equiv 5 \pmod{7}, \; 8 \equiv 1 \pmod{7}, \; 20 \equiv 6 \pmod{7}.$$

In general, if $a \equiv b \pmod{n}$ and k is a constant, then

$$a + k \equiv b + k \pmod{n}, \; a - k \equiv b - k \pmod{n}, \; ka \equiv kb \pmod{n}.$$

But we must be careful with division: for example, if we try to divide the congruence $15 \equiv 3 \pmod{6}$ by 3, we get $5 \equiv 1 \pmod{6}$, which is false. In general, we can divide congruences mod n by a constant k only when n and k are coprime: for example, we can divide the congruence $120 \equiv 15 \pmod{7}$ by 5 to give $24 \equiv 3 \pmod{7}$, because 5 and 7 are coprime. The general rule is:

If $ka \equiv kb \pmod{n}$ and if gcd $(k, n) = 1$, then
$a \equiv b \pmod{n}$.

But if k and n are not coprime, then we have to change the modulus:

If $ka \equiv kb \pmod{n}$ and if gcd $(k, n) = d$, then
$a \equiv b \pmod{n/d}$.

Several results from Chapter 2 can be conveniently stated in terms of congruences.

For example:

Every square has the form $4n$ or $4n + 1$, for some integer n

can be restated as

Every square is congruent to 0 or 1 (mod 4),

and other results on squares and cubes can be rewritten as

The square of every odd number is congruent to 1 (mod 8),
Every cube is congruent to 0, 1, or 8 (mod 9).

We can also revisit our earlier results on the divisibility of the integer

$$n = \left(a_k \times 10^k\right) + \left(a_{k-1} \times 10^{k-1}\right) + \cdots + \left(a_2 \times 10^2\right) + \left(a_1 \times 10^1\right) + \left(a_0 \times 10^0\right)$$

by various small numbers. For example:

Divisibility by 10: n is divisible by 10 if and only if its last digit is 0.

By the congruence rules, $n \equiv a_0 \pmod{10}$, which is congruent to 0 (mod 10) when a_0 is 0.

Divisibility by 4: n is divisible by 4 if and only if n ends in 00, 04, 08, ..., or 96:

Here, $n \equiv 10a_1 + a_0 \pmod{4}$, which is congruent to 0 (mod 4) when the two-digit number a_1a_0 is 00, 04, . . . , or 96.

Divisibility by 9: n is divisible by 9 if and only if the sum of its digits is divisible by 9:

Here, the powers of 10 are all congruent to 1 (mod 9), and so

$$n \equiv a_k + a_{k-1} + \cdots + a_1 + a_0 \pmod{9}.$$

The method of 'Casting out nines' works because every number is congruent to its digital root (mod 9).

Testing for Mersenne primes

In Chapter 3 we saw that some Mersenne numbers (such as $2^{127} - 1$) are prime, whereas others (such as $2^{67} - 1$) are not. A method for testing whether a given Mersenne number is prime was discovered by Edouard Lucas in 1876 and refined in the 1930s by D. H. Lehmer:

The Lucas–Lehmer test. Let $M_p = 2^p - 1$, where p is odd prime, and consider the sequence of numbers $s_0, s_1, s_2, s_3, \ldots$ in which $s_0 = 4$ and each successive number is defined by

$$s_{n+1} \equiv s_n^2 - 2 \pmod{M_p}.$$

Then M_p is prime if and only if the number s_{p-2} is congruent to 0 (mod M_p).

For example, if $p = 7$ and $M_p = 2^7 - 1 = 127$, then we have

$$s_0 = 4,$$
$$s_1 = 4^2 - 2 = 14 \equiv 14 \pmod{127},$$
$$s_2 = 14^2 - 2 = 194 \equiv 67 \pmod{127},$$
$$s_3 = 67^2 - 2 = 4487 \equiv 42 \pmod{127},$$
$$s_4 = 42^2 - 2 = 1762 \equiv 111 \pmod{127},$$

and $s_5 = 111^2 - 2 = 12{,}319$, which is congruent to 0 (mod 127), so 127 is prime.

But, if $p = 11$ and $M_p = 2^{11} - 1 = 2047$, then we have, after some calculation,

$$
\begin{aligned}
s_0 &= 4, \\
s_1 &= 4^2 - 2 \equiv 14 \pmod{2047}, \\
s_2 &= 14^2 - 2 \equiv 194 \pmod{2047}, \\
s_3 &= 194^2 - 2 \equiv 788 \pmod{2047}, \\
s_4 &= 788^2 - 2 \equiv 701 \pmod{2047}, \\
s_5 &= 701^2 - 2 \equiv 119 \pmod{2047}, \\
s_6 &= 119^2 - 2 \equiv 1877 \pmod{2047}, \\
s_7 &= 1877^2 - 2 \equiv 240 \pmod{2047}, \\
s_8 &= 240^2 - 2 \equiv 282 \pmod{2047},
\end{aligned}
$$

and $s_9 = 282^2 - 2 \equiv 1736$, which is not congruent to $0 \pmod{2047}$, so 2047 isn't prime.

Congruences and the calendar

The seven days of the week cycle around like the hours on a clock (see Figure 24).

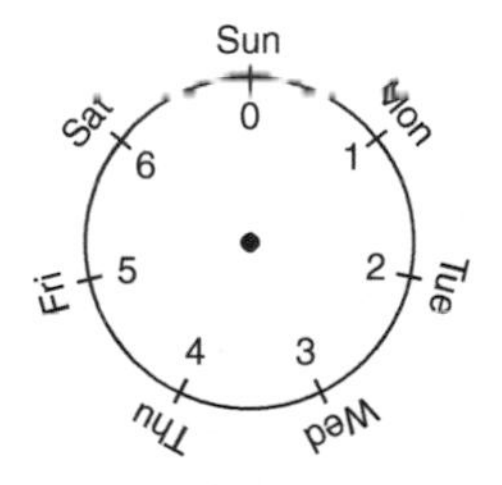

24. A 7-day clock.

If it's now Thursday, then in four days it will be Monday.

If it's now Saturday, then in three days it will be Tuesday.

The analogy is more obvious if we number the days of the week and work (mod 7):

Sunday = 0, Monday = 1, Tuesday = 2, Wednesday = 3,
Thursday = 4, Friday = 5, Saturday = 6.

The two statements above then become $4 + 4 \equiv 1 \pmod 7$ and $6 + 3 \equiv 2 \pmod 7$.

We'll use this analogy to answer various questions about the calendar: for example,

On which day of the week will 25 December 2099 fall?

In which years in the 21st century does February have five Sundays?

To simplify matters, we'll concentrate mainly on dates in the 21st century.

The Gregorian calendar, introduced by Pope Gregory XIII in 1582, has been in use in the UK and its former colonies since 1752. In this calendar a regular year has 365 days and a leap year has an extra 'leap day' on 29 February. The leap years are those that are divisible by 4, except for the century years that are not also divisible by 400: so 2000 and 2400 are leap years, but 1900 and 2100 are not.

Because $365 \equiv 1 \pmod 7$ and $366 \equiv 2 \pmod 7$, the day of the week on which any particular date falls advances by 1 each year, or by 2 when a leap day intervenes: for example, 1 January fell on a Monday in 2001, on a Tuesday in 2002, on a Wednesday in 2003, on a Thursday in 2004, and on a Saturday in 2005.

It's useful to note that the days of the calendar repeat every 28 years in which no century year intervenes: this is because the number of days (ordinary days and leap days) is

$$(28 \times 365) + 7,$$

which is divisible by 7. If we wish to take account of the century years, we can likewise check that the days of the calendar repeat every 400 years.

How can we calculate the day of the week on which a given date falls? Several puzzlers have discovered clever methods for doing so, including one by Gauss and a modern one called 'Doomsday' devised by the English mathematician John Conway. Here we present a method invented by the Oxford mathematician Charles L. Dodgson, better known as Lewis Carroll, the author of *Alice's Adventures in Wonderland*. He published it under his pen-name in March 1887, and claimed to be able to carry out all the mental calculations in about 20 seconds.

Carroll's method involves calculating four numbers—the century, year, month, and day numbers—and finding their sum (mod 7). We'll illustrate it by calculating the day of the week on which 25 December 2099 will fall.

Century number. Divide the first two digits of the year by 4, subtract the remainder from 3, and multiply the result by 2.

> So for the year 2099, we divide 20 by 4, giving a remainder of 0; subtracting this from 3 gives 3, and multiplying this by 2 gives the century number **6**.

Year number. Divide the last two digits of the year by 12, and add the quotient, the remainder, and the number of times that 4 divides into the remainder.

> So for the year 2099, we divide 99 by 12, giving a quotient of 8 and a remainder of 3; 4 divides this remainder no times, so the year number is $8 + 3 + 0 =$ **11**.

Month number. Carroll's method for finding the month number was somewhat complicated, so we omit it. It yields the following table:

January: 0 February: 3 March: 3 April: 6
May: 1 June: 4 July: 6 August: 2 September: 5
October: 0 November: 3 December: 5

So for 25 December, the month number is **5**.

Day number: this is simply the day of the month.

So for 25 December, the day number is **25**.

Finally, the total of the four numbers then must be reduced by 1 if the date is in January or February of a leap year.

So for 25 December 2099, which is not in a leap year, adding the four numbers gives

$$\mathbf{6} + \mathbf{11} + \mathbf{5} + \mathbf{25} = 47 \equiv \mathbf{5} \pmod{7},$$

so it will fall on a Friday.

We can now return to the following question that we asked earlier:

In which years in the 21st century does February have five Sundays?

For this to happen, the year must be a leap year, and the five Sundays must be 1, 8, 15, 22, and 29 February, so we need only to find out when 1 February is a Sunday. Now we know that 1 January 2001 was on a Monday, and that January has $31 \equiv 3 \pmod{7}$ days, so that 1 February 2001 was on a Thursday. It follows that 1 February was on a Friday in 2002, on a Saturday in 2003, and on a Sunday in 2004. So February 2004 had five Sundays. Because the cycle of days repeats every 28 years, February also has five Sundays in the leap years 2032, 2060, and 2088.

Solving linear congruences

An ancient Chinese puzzle concerns a band of pirates and a hoard of gold coins:

A band of 17 pirates stole a sack of gold coins.
When they tried to divide the fortune into equal portions, 4 coins remained.
In the ensuing brawl over who should get the extra coins, one pirate was killed.
The wealth was then redistributed, but this time an equal division left 10 coins over.
Again an argument developed in which another pirate was killed.
But now the total fortune could be evenly distributed among the survivors.
What was the least number of coins that could have been stolen?

If x is the total number of coins, we can write down three congruences:

at the first stage, with 17 pirates and 4 coins left over,

$$x \equiv 4 \pmod{17},$$

at the second stage, with 16 pirates and 10 coins left over,

$$x \equiv 10 \pmod{16},$$

at the third stage, with 15 pirates and no coins left over,

$$x \equiv 0 \pmod{15}.$$

Can we find a number x that simultaneously satisfies these three congruences?

We'll solve this puzzle later, but first we'll need to solve linear congruences in general: these have the form $ax \equiv b \pmod{n}$, where n is a positive integer, a and b are given integers, and our task is to find the unknown x. We'll then look at some problems that involve simultaneous linear congruences, such as the pirates puzzle above. Finally, we'll extend our explorations to some quadratic congruences of the form $x^2 \equiv b \pmod{n}$.

We'll start by looking in turn at three congruences:

$$5x \equiv 2 \pmod{6}, \quad 4x \equiv 2 \pmod{6}, \quad 4x \equiv 3 \pmod{6}.$$

For the first congruence we discover, after a little experimentation, that $x = 4$ is a solution, because

$$5 \times 4 = 20 \equiv 2 \pmod 6.$$

Exploring a little further, we find that $x = 10$ and $x = 16$ are also solutions, as are any other numbers that are congruent to $4 \pmod 6$. These turn out to be the only solutions.

The second congruence has two solutions, $x = 2$ and $x = 5$, or more generally any number that is congruent to 2 or 5 (mod 6).

The third congruence has no solutions.

To see why these different situations arise, let's begin with the third congruence.

If $4x \equiv 3 \pmod 6$, then the left-hand side is always even, and so can never equal a number of the form $3 \pmod 6$. So this congruence can have no solutions.

Turning our attention to the second congruence, we note that if $4x \equiv 2 \pmod 6$, then $2x \equiv 1 \pmod 3$, after cancelling throughout by 2. This has the single solution $x \equiv 2 \pmod 3$, which corresponds to $x \equiv 2$ or $5 \pmod 6$.

For the first congruence, $5x \equiv 2 \pmod 6$, we find, after some experimentation, that multiplication by 5 gives the congruence $25x \equiv 10 \pmod 6$—that is, $x \equiv 4 \pmod 6$. This is the only solution.

In general, we have the following rules for solving the congruence $ax \equiv b \pmod n$.

> If $\gcd(a, n) = 1$, then there is a single solution for $x \pmod n$.

If gcd $(a, n) = d$, and if d divides b, then there are d solutions (mod n), which correspond to the single solution of the congruence $(a/d)x \equiv b/d \pmod{n/d}$.

If gcd (a, n) doesn't divide b, then the congruence has no solutions.

For the congruences above:

when $5x \equiv 2 \pmod 6$, gcd $(5, 6) = 1$, and the only solution is $x \equiv 4 \pmod 6$;

when $4x \equiv 2 \pmod 6$, gcd $(4, 6) = 2$, so there are two solutions, $x \equiv 2$ and $5 \pmod 6$, which correspond to the single solution $x = 2 \pmod 3$ of the congruence $2x \equiv 1 \pmod 3$;

when $4x \equiv 3 \pmod 6$, gcd $(4, 6) = 2$, which doesn't divide 3, so there are no solutions.

Simultaneous linear congruences

An ancient puzzle, posed in the 4th century by the Chinese mathematician Sunzi, asks:

There are an unknown number of things. If we count by threes, there's a remainder of 2; if we count by fives, there's a remainder of 3; if we count by sevens, there's a remainder of 2. Find the number of things.

He answered this problem, together with many similar examples, in the *Sunzi suan jing* (Mathematical Classic of Master Sun), and over the next millennium it resurfaced in many lands and in many forms, such as the following:

I have some eggs. Arranged in rows of 3 there are 2 left over, in rows of 5 there are 3 left over, and in rows of 7 there are 2 left over. How many eggs have I altogether?

In either version, we seek a number x that simultaneously satisfies the congruences

$$x \equiv 2 \pmod 3), \quad x \equiv 3 \pmod 5), \quad \text{and} \quad x \equiv 2 \pmod 7).$$

We can answer Sunzi's problem by first trawling through the numbers congruent to 2 (mod 7) until we reach a number that's also congruent to 3 (mod 5): this gives

$$x = 2 \text{ (no)}, \quad 9 \text{ (no)}, \quad 16 \text{ (no)}, \quad 23 \text{ (yes)}.$$

So $x = 23$ is a solution of the last two congruences, and it also satisfies the first one. Other answers can then be found by adding multiples of $3 \times 5 \times 7 = 105$, because 105 is congruent to 0 (mod 3), 0 (mod 5), and 0 (mod 7): for example, $x = 128$ and $x = 253$ are also solutions to Sunzi's problem.

We were fortunate that a solution to the last two congruences also satisfies the first one, but this doesn't usually happen. For example, let's return to the pirates puzzle, where we seek a simultaneous solution to the congruences

$$x \equiv 4 \pmod{17}, \quad x \equiv 10 \pmod{16}, \quad x \equiv 0 \pmod{15}.$$

To solve this puzzle we first trawl through the numbers congruent to 4 (mod 17) until we reach a solution to the second congruence:

$$x = 4 \text{ (no)}, \quad 21 \text{ (no)}, \quad 38 \text{ (no)}, \quad 55 \text{ (no)}, \quad 72 \text{ (no)},$$
$$89 \text{ (no)}, \quad 106 \text{ (yes)}.$$

We next trawl through the simultaneous solutions of the first two congruences, by adding $17 \times 16 = 272$ each time, until we reach a solution of the third congruence:

$$x = 106 \text{ (no)}, \quad 378 \text{ (no)}, \quad 650 \text{ (no)}, \quad 922 \text{ (no)},$$
$$1194 \text{ (no)}, \quad 1466 \text{ (no)}, \quad 1738 \text{ (no)}, \quad 2010 \text{ (yes)}.$$

So $x = 2010$ satisfies all three congruences and is the smallest solution of the pirates problem. Other solutions can then be obtained by adding multiples of $17 \times 16 \times 15 = 4080$.

It turns out that any collection of simultaneous congruences has a single solution, provided that the moduli are coprime in pairs: for example, in Sunzi's problem, the moduli 3 and 5 are coprime, as are 3 and 7, and 5 and 7, and the same is true for the moduli in the pirates puzzle. A statement of what became known as the Chinese remainder theorem, in the case of three congruences, is as follows:

Chinese remainder theorem: Let n_1, n_2, and n_3 be positive integers with

$$\gcd\ (n_1, n_2) = \gcd\ (n_1, n_3) = \gcd\ (n_2, n_3) = 1,$$

and let $N = n_1 \times n_2 \times n_3$.

Then the linear congruences

$$x \equiv b_1 \pmod{n_1}, \quad x \equiv b_2 \pmod{n_2}, \quad x \equiv b_3 \pmod{n_3}$$

have a simultaneous solution which is unique (mod N).

We conclude this section with another poser that leads to solving simultaneous linear congruences. A number n is *self-replicating* if its square n^2 terminates in the digits of the original number n: for example, 25 is self-replicating because $25^2 = 6\underline{25}$, which terminates in 25. Are there any more?

The 1-digit self-replicating numbers are 0, 1, 5, and 6, because $0^2 = \underline{0}$, $1^2 = \underline{1}$, $5^2 = 2\underline{5}$, and $6^2 = 3\underline{6}$. These numbers all satisfy the congruence $n^2 \equiv n \pmod{10}$.

For 2-digit self-replicating numbers, n is self-replicating if $n^2 \equiv n \pmod{100}$. But if 100 divides $n^2 - n$, then so do 4 and 25, and we get the simultaneous congruences

$$n^2 \equiv n \pmod 4 \quad \text{and} \quad n^2 \equiv n \pmod{25}.$$

It follows from the first of these congruences that $n \equiv 0$ or $1 \pmod 4$.

For the second congruence, we notice that 25 divides $n^2 - n = n \times (n - 1)$, and so either 5 divides both n and $n - 1$ (which cannot happen) or 25 must divide n or $n - 1$. It follows that

$n \equiv 0$ or 1 (mod 25). So we have four pairs of simultaneous linear congruences:

$n \equiv 0 \pmod 4$ and $n \equiv 0 \pmod{25}$, with solution $n = 0$ which isn't a 2-digit number;

$n \equiv 1 \pmod 4$ and $n \equiv 0 \pmod{25}$: this has the solution $n = 25$, which we saw earlier;

$n \equiv 0 \pmod 4$ and $n \equiv 1 \pmod{25}$: this has the solution $n = 76$, with $76^2 = 57\underline{76}$;

$n \equiv 1 \pmod 4$ and $n \equiv 1 \pmod{25}$, with solution $n = 1$ which isn't a 2-digit number.

So the only two 2-digit self-replicating numbers are $n = 25$ and $n = 76$.

Likewise, for 3-digit self-replicating numbers, we have the congruence $n^2 \equiv n \pmod{1000}$, from which we get the simultaneous congruences

$$n^2 \equiv n \pmod 8 \quad \text{and} \quad n^2 \equiv n \pmod{125}.$$

These in turn lead to the simultaneous linear congruences $n \equiv 0$ or 1 (mod 8) and $n \equiv 0$ or 1 (mod 125), and on examining these in pairs, as above, we find that the only 3-digit self-replicating numbers are $n = 376$ and 625, with $376^2 = 141{,}\underline{376}$ and $625^2 = 390{,}\underline{625}$.

Squares and non-squares

We conclude this chapter by looking at some more quadratic congruences. Our explorations will lead us to one of the most important results in number theory, the *law of quadratic reciprocity*.

We'll start with the congruence

$$x^2 + 7x + 10 \equiv 0 \pmod{11}.$$

We can solve this by multiplying by 4 and adding 9 to both sides, giving

$$4x^2 + 28x + 49 \equiv 9 \pmod{11},$$

$$\text{so that} \quad (2x + 7)^2 \equiv 3^2 \pmod{11}.$$

Taking the square root now gives $2x + 7 \equiv 3 \pmod{11}$ or $2x + 7 \equiv -3 \pmod{11}$, and these two linear congruences can then be solved to give $x \equiv 9 \pmod{11}$ and $x \equiv 6 \pmod{11}$. These solutions are correct, because

$$9^2 + (7 \times 9) + 10 = 154 \equiv 0 \pmod{11}$$

$$6^2 + (7 \times 6) + 10 = 88 \equiv 0 \pmod{11}.$$

It turns out that many quadratic congruences can likewise be rewritten in the form $y^2 = b \pmod{n}$—here, $y = 2x + 7$, $b = 9$, and $n = 11$—and this leads us to investigate which numbers b are squares (mod n) and which ones are non-squares. We'll now explore this question when n is an odd prime number.

Let's first find the squares (mod p) when $p = 3, 5, 7$, and 11. Because $0 = 0^2$ is always a square, whatever the value of p, we'll seek only non-zero squares.

When $p = 3$, the non-zero squares are $1^2 = 1 \equiv 1 \pmod{3}$ and $2^2 = 4 \equiv 1 \pmod{3}$,

so the only non-zero square (mod 3) is 1, and the only non-square is 2.

When $p = 5$, the non-zero squares are

$$1^2 = 1 \equiv 1 \pmod{5}, \qquad 2^2 = 4 \equiv 4 \pmod{5},$$
$$3^2 = 9 \equiv 4 \pmod{5}, \qquad 4^2 = 16 \equiv 1 \pmod{5},$$

so the non-zero squares (mod 5) are 1 and 4, and the non-squares are 2 and 3.

When $p = 7$, the non-zero squares are:

$$\begin{aligned} 1^2 &= 1 \equiv 1 \pmod 7, & 2^2 &= 4 \equiv 4 \pmod 7, \\ 3^2 &= 9 \equiv 2 \pmod 7, & 4^2 &= 16 \equiv 2 \pmod 7, \\ 5^2 &= 25 \equiv 4 \pmod 7, & 6^2 &= 36 \equiv 1 \pmod 7, \end{aligned}$$

so the non-zero squares (mod 7) are 1, 2, and 4, and the non-squares are 3, 5, and 6.

When $p = 11$, the non-zero squares are:

$$\begin{aligned} 1^2 &= 1 \equiv 1 \pmod{11}, & 2^2 &= 4 \equiv 4 \pmod{11}, \\ 3^2 &= 9 \equiv 9 \pmod{11}, & 4^2 &= 16 \equiv 5 \pmod{11}, \\ 5^2 &= 25 \equiv 3 \pmod{11}, & 6^2 &= 36 \equiv 3 \pmod{11}, \\ 7^2 &= 49 \equiv 5 \pmod{11}, & 8^2 &= 64 \equiv 9 \pmod{11}, \\ 9^2 &= 81 \equiv 4 \pmod{11}, & 10^2 &= 100 \equiv 1 \pmod{11}, \end{aligned}$$

so the non-zero squares (mod 11) are 1, 3, 4, 5, and 9, and the non-squares are 2, 6, 7, 8, and 10.

We notice that in each case the numbers of squares and non-squares are the same, and this is the case for all primes. The squares are often called *quadratic residues* (mod p) and the non-squares are called *quadratic non-residues* (mod p).

If we multiply two squares (mod p) together, then the product is also a square: this is because

if $x \equiv a^2 \pmod p$ and $y \equiv b^2 \pmod p$,

then $xy \equiv a^2b^2 \equiv (ab)^2 \pmod p$.

We can likewise prove that

the product of any two non-squares (mod p) is a square,

and

the product of a square and a non-square (mod p) is a non-square.

For example, when $p = 11$,

6 and 7 are both non-squares, and their product $6 \times 7 = 42 \equiv 9 \pmod{11}$ is a square;

5 is a square and 7 is a non-square, and their product $5 \times 7 = 35 \equiv 2 \pmod{11}$ is a non-square.

Given an odd prime number p, how can we decide whether a given number a is a square or a non-square (mod p)? For example, knowing that 2027 is a prime, is 1066 a square or a non-square (mod 2027)? We'll answer this question later.

In 1798 Adrien-Marie Legendre introduced a useful notation that helps us when answering such questions. If p is an odd prime number and a is a number that's not divisible by p, then his *Legendre symbol* (a / p) is defined by

$$(a/p) = 1 \text{ if } a \text{ is a square } \pmod{p},$$

$$(a/p) = -1 \text{ if } a \text{ is a non-square } \pmod{p}:$$

for example, $(5/11) = 1$ and $(6/11) = (7/11) = -1$, because 5 is a square and 6 and 7 are non-squares (mod 11). We note that, for any odd prime number p, and any number a that is not divisible by p,

$$(a^2/p) = 1, \text{because } a^2 \text{ is always a square:}$$

for example, $(9/11) = (3^2/11) = 1$.

There are some useful rules that help us to find Legendre symbols. The first is:

If a and b are not divisible by p, and if $a \equiv b \pmod{p}$, then $(a/p) = (b/p)$.

For example, 42 is a square (mod 11), because $42 \equiv 9 \pmod{11}$, and so

$$(42/11) = (9/11) = 1.$$

The second rule restates the above remarks about multiplying squares and non-squares:

If a and b are not divisible by p, then $(ab/p) = (a/p) \times (b/p)$.

For example, 42 is a square (mod 11), because

$$(42/11) = (6/11) \times (7/11) = -1 \times -1 = 1.$$

It can also be shown that, for any odd prime p,

$$(-1/p) = 1 \text{ if } p \equiv 1 \pmod{4},$$
$$(-1/p) = -1 \text{ if } p \equiv 3 \pmod{4};$$

$$(2/p) = 1 \text{ if } p \equiv 1 \text{ or } 7 \pmod{8},$$
$$(2/p) = -1 \text{ if } p \equiv 3 \text{ or } 5 \pmod{8}.$$

These results conveniently tell us when -1 and 2 are squares (mod p). For example,

$(-1/17) = 1$ and $(-1/19) = -1$,
because $17 \equiv 1 \pmod{4}$ and $19 \equiv 3 \pmod{4}$;

$(2/17) = 1$ and $(2/19) = -1$,
because $17 \equiv 1 \pmod{8}$ and $19 \equiv 3 \pmod{8}$.

Suppose next that we wish to decide whether 41 is a square (mod 23).

By the first rule we can write

$$\begin{aligned} (41/23) &= (18/23) && \text{because } 41 \equiv 18 \pmod{23} \\ &= (2/23) \times (9/23) && \text{because } 18 = 2 \times 9 \\ &= 1 \times 1 && \text{because } 23 \equiv 7 \pmod{8} \text{ and} \\ & && 9 \text{ is a square.} \end{aligned}$$

So $(41/23) = 1$, and 41 is a square (mod 23).

Another way to deduce this is to spot that $41 \equiv 64 = 8^2 \pmod{23}$.

Suppose now that we wish to decide whether 42 is a square (mod 23)?

Here, $42 \equiv 19 \pmod{23}$, and so $(42/23) = (19/23)$. But how can we decide whether 19 is a square (mod 23) without having to calculate all the squares (mod 23)?

To answer this, we'll introduce the 'law of quadratic reciprocity', which provides a reciprocal relationship between the squares (mod p) and the squares (mod q) when p and q are both odd primes. It was known to both Euler and Lagrange, but it was Gauss who presented no fewer than eight proofs of it—indeed, he became so excited by it that he called it his 'Golden theorem'. There are now around two hundred proofs of the reciprocity law, which states:

Law of quadratic reciprocity: If p and q are odd primes, then

$$(p/q) = (q/p) \quad \text{if either } p \text{ or } q \equiv 1 \pmod 4,$$
$$(p/q) = -(q/p) \quad \text{if both } p \text{ and } q \equiv 3 \pmod 4.$$

For example,

$$(13/23) = (23/13), \quad \text{because } 13 \equiv 1 \pmod 4,$$
$$(11/23) = -(23/11), \quad \text{because both 11 and } 23 \equiv 3 \pmod 4.$$

We can now return to our problem of deciding whether 42 is a square (mod 23).

Above we saw that $(42/23) = (19/23)$. But

$$\begin{aligned} (19/23) &= -(23/19) && \text{because 19 and } 23 \equiv 3 \pmod 4 \\ &= -(4/19) && \text{because } 23 \equiv 4 \pmod{19} \\ &= -1 && \text{because 4 is a square.} \end{aligned}$$

So $(42/23) = -1$, and 42 is a non-square (mod 23).

We conclude this chapter by returning to our earlier problem of deciding whether 1066 is a square or a non-square (mod 2027). We can now answer this by applying the law of quadratic reciprocity several times to reduce the numbers involved. Because $1066 = 2 \times 13 \times 41$, we have

$$(1066/2027) = (2/2027) \times (13/2027) \times (41/2027).$$

But $(2/2027) = -1$, because $2027 \equiv 3 \pmod 8$. Also,

$$\begin{aligned}
(13/2027) &= (2027/13) && \text{because } 13 \equiv 1 \pmod 4 \\
&= (-1/13) && \text{because } 2027 \equiv -1 \pmod{13} \\
&= 1 && \text{because } 13 \equiv 1 \pmod 4.
\end{aligned}$$

Finally,

$$\begin{aligned}
(41/2027) &= (2027/41) && \text{because } 41 \equiv 1 \pmod 4 \\
&= (18/41) && \text{because } 2027 \equiv 18 \pmod{41} \\
&= (2/41) \times (9/41) && \\
&= 1 \times 1 && \text{because } 41 \equiv 1 \pmod 8 \\
& && \text{and 9 is a square} \\
&= 1.
\end{aligned}$$

Combining all of these results gives $(1066/2027) = -1 \times 1 \times 1 = -1$, and so 1066 is not a square (mod 2027).

In Chapter 6 we'll continue our explorations into congruences with some fundamental results of Fermat and Euler and with some applications to cryptography and to the shuffling of cards and the colouring of necklaces.

Chapter 5
More triangles and squares

There are many intriguing problems that can be expressed in terms of equations requiring whole-number solutions: such equations are called *Diophantine equations*. They are named after Diophantus of Alexandria who, as we mentioned in Chapter 1, wrote a classic text called *Arithmetica* which contained many questions with such solutions. In this chapter we'll present a number of Diophantine problems, ranging from finding right-angled triangles with integer sides to writing numbers as the sum of squares and higher powers, and we'll conclude with a brief account of one of number theory's most celebrated achievements—the proof of Fermat's so-called 'last theorem'.

Linear Diophantine equations

As an example of a Diophantine equation, we'll consider the equation

$$2x + 3y = 13.$$

If x and y can take any values, then there are infinitely many solutions—choose any number x, and calculate $y = (13 - 2x)/3$: for example, if $x = 1$, then $y = 11/3$.

If we now require x and y to be integers, then there are still infinitely many solutions—reducing the equation (mod 3) gives $2x \equiv 1 \pmod{3}$, so $x \equiv 2 \pmod{3}$ and the solutions become $x = 2 + 3k, y = 3 - 2k$, where k is an integer: for example, taking $k = 10$ gives the solution $x = 32, y = -17$.

But if we further require x and y to be *positive* integers, then there are only two solutions:

$$x = 2,\ \ y = 3\,(\text{for } k = 0) \quad \text{and} \quad x = 5,\ \ y = 1\,(\text{for } k = 1)\,.$$

These solutions are illustrated in Figure 25.

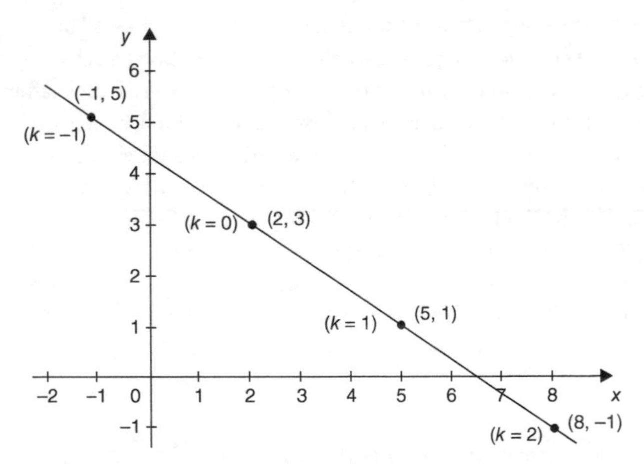

25. Some solutions of the Diophantine equation $2x + 3y = 13$.

In general, it can be shown that:

If a, b, and c are given integers, then the linear equation

$$ax + by = c$$

has a solution in integers if and only if gcd (a, b) divides c.

For the example above, we had $a = 2$, $b = 3$, and gcd $(2, 3) = 1$, which divides 13.

Moreover:

If gcd $(a, b) = 1$, and if we can spot a particular solution, $x = X, y = Y$, then every solution can be written in the form

$$x = X + bk, \quad y = Y - ak, \quad \text{where } k \text{ is an integer.}$$

For the example above, $X = 2$ and $Y = 3$, and gcd $(a, b) = 1$, and so the solutions all have the form

$$x = 2 + 3k, \quad y = 3 - 2k,$$

as we saw earlier.

Another type of linear Diophantine problem is the following, adapted from one of Leonardo Fibonacci in 1202:

If I can buy partridges for 3 cents, pigeons for 2 cents, and 2 sparrows for a cent, and if I spend 30 cents on buying 30 birds, how many birds of each kind must I buy?

Because the birds are indivisible, we seek whole-number solutions. Further, we'll assume that I buy at least one of each type of bird, so we seek *positive* solutions. If I buy a partridges, b pigeons, and c sparrows, then we can write down two equations:

$$\begin{aligned} a + b + c &= 30 \quad \text{for the total number of birds,} \\ 3a + 2b + c/2 &= 30 \quad \text{for the total number of cents.} \end{aligned}$$

At first sight it may seem that, with three unknowns and only two equations, we cannot answer the question. But we have a further piece of information: a, b, and c are all positive integers. Let's proceed:

Multiplying the second equation by 2 gives

$$6a + 4b + c = 60,$$

and then on subtracting the first equation from this to eliminate c, we get

$$5a + 3b = 30, \text{so} \quad 3b = 5\,(6 - a)\,.$$

So b is divisible by 5, and cannot be 10 (because a would then be 0). So $b = 5$, giving $a = 3$ and (from the first equation) $c = 22$. So I should buy 3 partridges, 5 pigeons, and 22 sparrows.

Right-angled triangles

In Chapter 1 we saw that the sides a, b, and c of a right-angled triangle satisfy the Pythagorean theorem,

$$a^2 + b^2 = c^2,$$

and we gave two examples of right-angled triangles with integer sides:

$$3^2 + 4^2 = 5^2 \text{ and } 5^2 + 12^2 = 13^2.$$

We call (a, b, c) a *Pythagorean triple* if $a^2 + b^2 = c^2$ and a, b, and c are positive integers, so (3, 4, 5) and (5, 12, 13) are Pythagorean triples. Our aim is to find all such triples.

We first note that if (a, b, c) is a Pythagorean triple, with $a^2 + b^2 = c^2$, then so is any multiple (ka, kb, kc), where k is a positive integer, because

$$(ka)^2 + (kb)^2 = k^2 \times (a^2 + b^2) = k^2 \times c^2 = (kc)^2:$$

for example, (30, 40, 50) is also a Pythagorean triple. Geometrically, multiples of a Pythagorean triple correspond to scalings of the right-angled triangle.

Such scalings are not very interesting, and so we shall mainly consider triples in which the sides have no common factor k, other than 1. We call these *primitive triples*: for example, (3, 4, 5) and (5, 12, 13) are both primitive triples. Can we find a formula for generating all primitive triples?

We first note that if (a, b, c) is a primitive triple, then any two of the numbers a, b, and c must be coprime, for if (for example) a and

b have a common factor that's greater than 1, then they must have a common prime factor p. It follows that p must divide $a^2 + b^2$, which equals c^2, and so p (being prime) also divides c. This contradicts the fact that a, b, and c have no common factor. We can therefore assume that

$$\gcd\ (a, b) = \gcd\ (a, c) = \gcd\ (b, c) = 1.$$

But we can say more. If (a, b, c) is a primitive triple, then a and b cannot both be even, as we've just seen. Can they both be odd? As we saw in Chapter 2, every square has the form $4n$ or $4n + 1$, and so if a and b are both odd, then a^2 and b^2 must have the form $4n + 1$, and c^2 must have the form $4n + 2$, which is impossible. So one of a and b must be even and the other must be odd, and c^2, and hence c, must also be odd. For definiteness, we'll always take a to be odd and b to be even.

Now, because a and c are both odd, their sum and difference are both even and we can write $c + a = 2u$ and $c - a = 2v$, for some integers u and v, with $u > v$. So $u + v = c$ and $u - v = a$.

Also, b is even and $b^2 = c^2 - a^2 = (c + a)(c - a)$, and so

$$(b/2)^2 = (c + a)/2 \times (c - a)/2 = u \times v.$$

What can we say about u and v? If $\gcd\ (u, v) = d$, where $d > 1$, then d divides $c + a$ and $c - a$, and so divides both c and a, which cannot happen. So u and v are coprime. Also, because uv is a square and u and v are coprime, u and v must separately be squares, and so we can write $u = x^2$ and $v = y^2$, for some integers x and y with $x > y$. So

$$c = u + v = x^2 + y^2, \quad a = u - v = x^2 - y^2,$$
$$\text{and}\ \ b^2 = 4uv = 4x^2y^2, \text{so } b = 2xy.$$

It can be checked that one of the numbers x and y is odd and the other is even, and that they are coprime. Tying all this together, we have our desired formula for primitive triples:

Primitive Pythagorean triples: If (a, b, c) is a primitive Pythagorean triple, then

$$a = x^2 - y^2, \quad b = 2xy, \quad c = x^2 + y^2,$$

where x and y are coprime integers, one odd and the other even, with $x > y$.

For example, if $x = 2$ and $y = 1$, then $a = 3$, $b = 4$, and $c = 5$, giving the triple (3, 4, 5). Likewise, if $x = 3$ and $y = 2$, then $a = 5$, $b = 12$, and $c = 13$, giving the triple (5, 12, 13).

We can use this recipe to draw up a table of primitive Pythagorean triples, by listing all pairs of coprime integers x and y with one odd and the other even and $x > y$, and calculating the corresponding values of a, b, and c. Table 4 lists all the primitive triples with no numbers exceeding 100. By extending this table as far as necessary, and then taking multiples, we can generate all the right-angled triangles with whole number sides.

Table 4. Primitive Pythagorean triples

x	y	:	a	b	c	x	y	:	a	b	c
2	1	:	3	4	5	7	2	:	45	28	53
3	2	:	5	12	13	7	4	:	33	56	65
4	1	:	15	8	17	7	6	:	13	84	85
4	3	:	7	24	25	8	1	:	63	16	65
5	2	:	21	20	29	8	3	:	55	48	73
5	4	:	9	40	41	8	5	:	39	80	89
6	1	:	35	12	37	9	2	:	77	36	85
6	5	:	11	60	61	9	4	:	65	72	97

We can also use our formula for primitive triples to answer such questions as:

How many primitive triples include the number 60?

Because 60 is even, we have $2xy = 60$, and so $xy = 30$. Remembering that $x > y$ and that x and y are coprime with one of them even and the other odd, we find the following four possibilities:

$x = 6$ and $y = 5$, giving the primitive triple $(11, 60, 61)$,
$x = 10$ and $y = 3$, giving the primitive triple $(91, 60, 109)$,
$x = 15$ and $y = 2$, giving the primitive triple $(221, 60, 229)$,
$x = 30$ and $y = 1$, giving the primitive triple $(899, 60, 901)$.

A similar question is:

How many right-angled triangles with whole-number sides have a side of length 29?

Because 29 is prime, the lengths of the sides must form a primitive triple, so either

$$29 = x^2 + y^2 \text{ or } 29 = x^2 - y^2, \text{ for some integers } x \text{ and } y.$$

In the former case, $x = 5$ and $y = 2$, and the triple $(x^2 - y^2, 2xy, x^2 + y^2)$ is $(21, 20, 29)$.

In the latter case, $29 = (x + y) \times (x - y)$, so $x + y = 29$ and $x - y = 1$. So $x = 15$ and $y = 14$, and the triple $(x^2 - y^2, 2xy, x^2 + y^2)$ is $(29, 420, 421)$.

Sums of squares

Having just investigated the equation $a^2 + b^2 = c^2$, we may ask a more general question that can be traced back to Diophantus:

Which numbers can be written as the sum of two perfect squares?

The first few examples are:

$$1 = 1^2 + 0^2, \quad 2 = 1^2 + 1^2, \quad 4 = 2^2 + 0^2, \quad 5 = 2^2 + 1^2,$$
$$8 = 2^2 + 2^2, \quad 9 = 3^2 + 0^2, \quad 10 = 3^2 + 1^2, \quad 13 = 3^2 + 2^2,$$
$$16 = 4^2 + 0^2, \quad 17 = 4^2 + 1^2, \quad 18 = 3^2 + 3^2, \quad 20 = 4^2 + 2^2.$$

The numbers up to 20 that cannot be written as the sum of two squares are 3, 6, 7, 11, 12, 14, 15, and 19. Can we deduce a general pattern from this?

The first thing to remember is that any square has the form $4n$ or $4n + 1$, and so the sum of two squares must have the form $4n$, $4n + 1$, or $4n + 2$. So any number of the form $4n + 3$ cannot be a sum of two squares, and this rules out 3, 7, 11, 15, and 19. We also note that 6, 12, and 14, which are multiples of the forbidden numbers 3 and 7, cannot be written as the sum of two squares, whereas 9 and 18, which are multiples of 3^2, can be so written. By using these observations as guidelines, it turns out that we can completely describe which numbers can be written as the sum of two squares. The following result was stated by Fermat, and proved by Legendre in 1798:

> *Sum of two squares*: A number can be written as the sum of two squares if and only if every prime factor that is congruent to 3 (mod 4) occurs to an even power.

For example, $117 = 3^2 \times 13$ is the sum of two squares $(9^2 + 6^2)$, because 3 occurs to an even power, whereas $120 = 2^3 \times 3 \times 5$ and $135 = 3^3 \times 5$ cannot be the sum of two squares because 3 occurs to an odd power in each case.

A useful result is that if m and n can be written as the sum of two squares, then so can their product mn. This is because, for $m = a^2 + b^2$ and $n = x^2 + y^2$, a little algebra confirms the multiplication rule:

$$mn = (a^2 + b^2) \times (x^2 + y^2) = (ax + by)^2 + (ay - bx)^2.$$

For example, knowing that $5 = 2^2 + 1^2$ and $13 = 3^2 + 2^2$, we can take $a = 2$, $b = 1$, $x = 3$, and $y = 2$, giving

$$\begin{aligned} 65 = 5 \times 13 &= \{(2 \times 3) + (1 \times 2)\}^2 + \{(2 \times 2) - (1 \times 3)\}^2 \\ &= (6 + 2)^2 + (4 - 3)^2 = 8^2 + 1^2, \end{aligned}$$

or, on rewriting $13 = 2^2 + 3^2$, we could take $a = 2$, $b = 1$, $x = 2$, and $y = 3$, giving

$$65 = 5 \times 13 = \{(2 \times 2) + (1 \times 3)\}^2 + \{(2 \times 3) - (1 \times 2)\}^2$$
$$= (4 + 3)^2 + (6 - 2)^2 = 7^2 + 4^2.$$

So some numbers can be written as the sum of two squares in more than one way: another example is

$$1105 = 33^2 + 4^2 = 32^2 + 9^2 = 24^2 + 23^2.$$

It follows from the above multiplication rule that if we can first decide which *prime numbers* can be written as the sum of two squares, then by combining them we can work out which numbers in general can be written in this way. This is another instance of the use of prime numbers as the building blocks for numbers in general. We now proceed to investigate the representation of prime numbers as the sum of two squares.

We have seen that $2 = 1^2 + 1^2$, and that the odd primes of the form $4n + 3$ cannot be written as the sum of two squares. But what about the odd primes of the form $4n + 1$? We can certainly write:

$$5 = 2^2 + 1^2, \quad 13 = 3^2 + 2^2, \quad 17 = 4^2 + 1^2, \quad 29 = 5^2 + 2^2,$$
$$37 = 6^2 + 1^2, \quad 41 = 5^2 + 2^2, \ldots.$$

This list illustrates the following general rule, which was first stated by Albert Girard in 1625 and in a letter from Fermat to Mersenne on Christmas Day 1640, and was eventually proved by Euler in 1754:

> An odd prime number p can be written as the sum of two squares if and only if p has the form $4n + 1$.
> Moreover, this can be done in only one way, apart from the order in which the two squares appear.

Before leaving this section, we ask a related question:

Which numbers can be written as the difference of two perfect squares?

This question is much easier to answer. Let's look at some examples:

$$1 = 1^2 - 0^2, \quad 3 = 2^2 - 1^2, \quad 4 = 2^2 - 0^2, \quad 5 = 3^2 - 2^2,$$
$$7 = 4^2 - 3^2, \quad 8 = 3^2 - 1^2, \quad 9 = 3^2 - 0^2, \quad 11 = 6^2 - 5^2,$$
$$12 = 4^2 - 2^2, \ldots .$$

But 2, 6, 10, ... cannot be written in this way, and it seems as though the following general rule applies:

Difference of two squares: A number can be written as the difference of two squares except when it has the form $4n + 2$.

To see why this is, we note again that every perfect square has the form $4n$ or $4n + 1$, and so the difference between two squares must have the form $4n$, $4n + 1$, or $4n + 3$. So the difference between two squares can never be of the form $4n + 2$. Moreover,

if $N = 4n$, then we can write

$$(n+1)^2 - (n-1)^2 = (n^2 + 2n + 1) - (n^2 - 2n + 1)$$
$$= 4n = N,$$

if $N = 4n + 1$, then we can write

$$(2n+1)^2 - (2n)^2 = (4n^2 + 4n + 1) - 4n^2 = 4n + 1 = N,$$

and if $N = 4n + 3$, then we can write

$$(2n+2)^2 - (2n+1)^2 = (4n^2 + 8n + 4) - (4n^2 + 4n + 1)$$
$$= 4n + 3 = N.$$

So every number can indeed be written as the difference of two squares, except when it has the form $4n + 2$.

Sums of more squares

We've seen that we cannot write every number as the sum of just two squares. Can we write every number as the sum of three squares (zero being allowed)? Certainly we have

$$3 = 1^2 + 1^2 + 1^2, \quad 6 = 2^2 + 1^2 + 1^2, \quad 11 = 3^2 + 1^2 + 1^2,$$
$$12 = 2^2 + 2^2 + 2^2, \quad 14 = 3^2 + 2^2 + 1^2, \quad 19 = 3^2 + 3^2 + 1^2.$$

But we cannot write 7 or 15 as the sum of only three squares. To see why, we note that every odd square has the form $8n + 1$ (see Chapter 2), and that every even square, being divisible by 4, has the form $8n$ or $8n + 4$. But we cannot combine three numbers of the form $8n$, $8n + 1$, or $8n + 4$ to give a number of the form $8n + 7$. Diophantus had claimed this result 1700 years ago, and Fermat proposed the following more general result, which was proved by Legendre in 1798:

> *Sum of three squares*: Every number can be written as the sum of three squares, except when it has the form $4^m \times (8k + 7)$, for some integers m and k.

For example,

$$7 = 4^0 \times 7, \quad 15 = 4^0 \times 15, \quad \text{and} \quad 23 = 4^0 \times 23$$

cannot be written as the sum of three squares, and nor can their multiples,

$$28 = 4^1 \times 7, \quad 240 = 4^2 \times 15, \quad 368 = 4^3 \times 23.$$

But these numbers can all be written as the sum of four squares: for example,

$$7 = 2^2 + 1^2 + 1^2 + 1^2, \qquad 15 = 3^2 + 2^2 + 1^2 + 1^2,$$
$$23 = 3^2 + 3^2 + 2^2 + 1^2, \qquad 28 = 5^2 + 1^2 + 1^2 + 1^2,$$
$$240 = 14^2 + 6^2 + 2^2 + 2^2, \quad 368 = 18^2 + 6^2 + 2^2 + 2^2.$$

In fact, every positive integer without exception can be written as the sum of four squares. This remarkable result was stated by Claude Bachet de Méziriac in 1621, and was proved by Joseph-Louis Lagrange using ideas of Euler:

> *Lagrange's four-square theorem*: Every number can be written as the sum of four squares.

As with Fermat's earlier result on the sum of two squares, it is enough to prove this for prime numbers only, and then to use a multiplication rule to obtain the general result. This rule, discovered by Euler, states that if m and n can each be written as the sum of four squares, then so can their product mn. This is because, for $m = a^2 + b^2 + c^2 + d^2$ and $n = w^2 + x^2 + y^2 + z^2$, we have

$$mn = (a^2 + b^2 + c^2 + d^2) \times (w^2 + x^2 + y^2 + z^2)$$
$$= (aw - bx - cy - dz)^2 + (bw + ax - dy + cz)^2$$
$$+ (cw + dx + ay - bz)^2 + (dw - cx + by + az)^2.$$

For example, knowing that $15 = 3^2 + 2^2 + 1^2 + 1^2$ and $23 = 3^2 + 3^2 + 2^2 + 1^2$, we can take $a = 3, b = 2, c = 1, d = 1, w = 3, x = 3, y = 2$, and $z = 1$, and write $345 = 15 \times 23$ as a sum of four squares as follows:

$$345 = 15 \times 23 = \{(3 \times 3) - (2 \times 3) - (1 \times 2) - (1 \times 1)\}^2$$
$$+ \{(2 \times 3) + (3 \times 3) - (1 \times 2) + (1 \times 1)\}^2$$
$$+ \{(1 \times 3) + (1 \times 3) + (3 \times 2) - (2 \times 1)\}^2$$
$$+ \{(1 \times 3) - (1 \times 3) + (2 \times 2) + (3 \times 1)\}^2$$
$$= (9 - 6 - 2 - 1)^2 + (6 + 9 - 2 + 1)^2$$
$$+ (3 + 3 + 6 - 2)^2 + (3 - 3 + 4 + 3)^2$$
$$= 0^2 + 14^2 + 10^2 + 7^2.$$

Higher powers

Having looked at sums of squares, we now turn our attention to sums of cubes and higher powers. A well-known story concerns the Cambridge mathematician G. H. Hardy who in 1918 was visiting his friend and colleague Srinivasa Ramanujan, one of the most intuitive mathematicians of all time, who was seriously ill in hospital. Together they'd solved some important problems in number theory. Hardy, finding it difficult to think of anything to say, remarked that his taxicab to the hospital had the number 1729 which seemed to be rather a dull number. As Hardy recalled, Ramanujan replied: 'No, Hardy! It is a very interesting number. It is the smallest number expressible as the sum of two cubes in two different ways.'
(These are $1000 + 729 = 10^3 + 9^3$ and $1728 + 1 = 12^3 + 1^3$.)

Waring's problem

As we have seen, Lagrange's four-square theorem asserts that every number can be written as the sum of four squares. In 1770 the Cambridge mathematician Edward Waring suggested that there were similar results for higher powers, claiming the following results for cubes and fourth powers:

> Every number can be written as the sum of nine cubes.
>
> Every number can be written as the sum of nineteen fourth powers.

Waring's claims were correct, and cannot be improved because there are numbers that actually require nine cubes and nineteen fourth powers, such as

$$\text{for cubes: } 23 = 2^3 + 2^3 + 1^3 + 1^3 + 1^3 + 1^3 + 1^3 + 1^3 + 1^3,$$
$$\text{for fourth powers: } 79 = 2^4 + 2^4 + 2^4 + 2^4 + 1^4 + 1^4 + 1^4 + 1^4 + 1^4 + 1^4 + 1^4 + 1^4 + 1^4 + 1^4 + 1^4 + 1^4 + 1^4 + 1^4 + 1^4.$$

Waring asked whether these ideas can be extended to higher powers:

Waring's problem: For each positive integer k, does there exist a number $g(k)$ such that every number can be written as the sum of $g(k)$ kth powers?

For example, $g(2) = 4$, by Lagrange's four-square theorem, and the results just asserted tell us that $g(3) = 9$ and $g(4) = 19$. Other known values are $g(5) = 37$ for sums of fifth powers, and $g(6) = 73$ for sums of sixth powers. In 1909, the German mathematician David Hilbert answered Waring's question in the affirmative by proving that there is such a number $g(k)$ for each value of k.

What can we say about $g(k)$? Around 1772, Johann Albrecht Euler, Leonhard's eldest son, suggested that

$$g(k) \geq 2^k + \lfloor 1.5^k \rfloor - 2, \quad \text{for all values of } k,$$

where $\lfloor x \rfloor$ is the integer part of x: for example, $\lfloor 1.5^3 \rfloor = \lfloor 3.375 \rfloor = 3$.

Note that,

$$\text{for } k = 3, \quad g(k) \geq 2^3 + [1.5]^3 - 2 = 8 + 3 - 2 = 9$$
$$\text{for } k = 4, \quad g(k) \geq 2^4 + [1.5]^4 - 2 = 16 + 5 - 2 = 19,$$

so this bound gives the correct value for $g(k)$ in these two cases. It has since been proved to give the right answer for all values of k up to 471.6 million, and with all this evidence it's believed to be correct in every case, although this has never been proved.

But we can say more. When $k = 3$, it turns out that only two numbers (23 and 239) actually need nine cubes, and that only a finite number need eight cubes. So we can say that,

> From some point onwards, every number can be written as the sum of just seven cubes.

It is also believed, although this has never been proved, that the number of cubes can be reduced further, possibly even to four.

For fourth powers it can be proved that,

> From some point onwards, every number can be written as the sum of just sixteen fourth powers.

Corresponding results can be proved for higher powers.

Fermat's last theorem

We conclude this chapter with a brief account of one of the most famous achievements in number theory—the proof of Fermat's last theorem.

In our discussion of Pythagorean triples, we saw how to find integer solutions of the Diophantine equation

$$a^2 + b^2 = c^2.$$

Can we likewise find integer solutions to the equations

$$a^3 + b^3 = c^3, \quad a^4 + b^4 = c^4, \quad a^5 + b^5 = c^5,$$

and in general,

$$a^n + b^n = c^n, \quad \text{for all } n \geq 3?$$

We can make a couple of general observations.

First, there are always 'trivial' solutions, in which one of the numbers a, b, and c is 0: for example,

$$5^3 + (-5)^3 = 0^4 \quad \text{and} \quad 2^4 + 0^4 = 2^4.$$

In what follows we're interested only in positive solutions.

Secondly, suppose that we knew that there were no integer solutions of the equation $a^3 + b^3 = c^3$. Then we could deduce that the same is true when the exponent is any multiple of 3. For example, the equation $x^{12} + y^{12} = z^{12}$ could then have no integer solutions, because we can rewrite it as

$$(x^4)^3 + (y^4)^3 = (z^4)^3,$$

$$\text{or } a^3 + b^3 = c^3 \quad (\text{where } a = x^4, b = y^4, \text{ and } c = z^4),$$

which we've assumed to have no solutions. So, when investigating solutions of the equation $a^n + b^n = c^n$, we can restrict our attention to the cases when n is 4 or a prime number.

Fermat became interested in the problem while reading a Latin translation of Diophantus's *Arithmetica* that had been published by Claude Bachet de la Méziriac in 1621 (see Figure 26). In the section on Pythagorean triples Fermat added a now-famous marginal comment:

> On the other hand, it is impossible for a cube to be written as a sum of two cubes or a fourth power to be written as a sum of two fourth powers or, in general, for any number

DIOPHANTI
ALEXANDRINI
ARITHMETICORVM
LIBRI SEX,
ET DE NVMERIS MVLTANGVLIS
LIBER VNVS.

Nunc primùm Græcè & Latinè editi, atque absolutissimis Commentariis illustrati.

AVCTORE CLAVDIO GASPARE BACHETO
MEZIRIACO SEBVSIANO, V.C.

LVTETIAE PARISIORVM,
Sumptibus SEBASTIANI CRAMOISY, via Iacobæa, ſub Ciconiis.
M. DC. XXI.
CVM PRIVILEGIO REGIS.

26. Bachet's translation of Diophantus's *Arithmetica*.

> which is a power greater than the second to be written as a sum of two like powers. I have a truly marvellous proof of this proposition which this margin is too narrow to contain.

This is often called 'Fermat's last theorem', because it became the last of Fermat's assertions to be proved—but it should perhaps have been called 'Fermat's conjecture':

> *Fermat's last theorem.* The equation $x^n + y^n = z^n$ has no positive integer solutions when $n \geq 3$.

Most mathematicians believe that Fermat couldn't have had a valid proof of his conjecture. Many attempts have since been made to find one, and the belief is that, if he really had found a correct argument, then it would surely have been rediscovered over the ensuing years.

Fermat himself proved that the equation $a^4 + b^4 = c^4$ can have no integer solutions. To do so, he invented a method of proof known as the *method of infinite descent*, showing by an algebraic argument that if this equation (or, more precisely, one closely related to it) had a solution in positive integers, then there would be another solution that involved smaller numbers than before. By repeating this argument over and over again, he'd then get smaller and smaller positive solutions—but this cannot happen indefinitely. This contradiction shows that the original solution couldn't have existed in the first place.

More than a century later, in 1770, Euler produced a proof for the case $n = 3$, but it had a gap in it that was subsequently filled by Legendre.

Major advances were made in the 1820s by Sophie Germain, a self-taught mathematician who made several substantial contributions to number theory and to the theory of elasticity. Interested in primes p for which $2p + 1$ is also prime (such as 2, 3, 5, 11, and 23), she proved several results concerning them—for example, that if $a^p + b^p = c^p$, then one of a, b, and c must be divisible by $2p + 1$, and one of them must be divisible by p^2. Then Legendre and Lejeune Dirichlet proved the theorem when $n = 5$, using her results, and in 1835 Gabriel Lamé of Paris proved it for $n = 7$.

A big advance was made by Ernst Kummer, who proved the result when n is a so-called *regular prime*: this settled the argument for almost all values of n that are less than 100. Over the years more and more cases were settled, and by the mid-20th century Fermat's conjecture had been proved for all values of n up to 2500, and by 1990 for all values up to 4 million. But there was still a long way to go!

In June 1993 Andrew Wiles, a British mathematician working at Princeton University in the US, announced to great excitement at a conference in Cambridge that he had proved Fermat's last theorem. Although this turned out to be premature, as a gap was found in his argument, he eventually patched it up with the help of his former student, Richard Taylor. Wiles's remarkable proof was published in 1995 to great acclaim. Fermat's last theorem had at last been proved (see Figure 27).

27. A postage stamp celebrates Andrew Wiles's proof of Fermat's last theorem.

Chapter 6
From cards to cryptography

In this chapter we introduce a fundamental theorem of Pierre de Fermat and apply it to a recreational problem on the shuffling of cards. We then present a generalization of Fermat's result that's due to Leonhard Euler, and a method of factorizing numbers that's also due to Fermat. We conclude by applying Euler's theorem to the sending of secret messages in cryptography, describing its use in ensuring the security of our credit cards.

Fermat's little theorem

Over two thousand years ago, Chinese mathematicians supposedly observed that

$$\begin{aligned}&2 \text{ divides } 2^2 - 2\ (= 2), \quad 3 \text{ divides } 2^3 - 2\ (= 6),\\ &5 \text{ divides } 2^5 - 2\ (= 30), \quad \text{and} \quad 7 \text{ divides } 2^7 - 2\ (= 126),\\ &\text{but } 4 \text{ doesn't divide } 2^4 - 2\ (= 14)\\ &\text{and } 6 \text{ doesn't divide } 2^6 - 2\ (= 62).\end{aligned}$$

This led them to claim that n divides $2^n - 2$ if and only if n is prime. Is their claim true? In this section we'll investigate this and similar questions.

In Chapter 4 we studied squares (mod p), where p is an odd prime, and we'll now look at higher powers. We'll start with the first few powers of 3 and 4 (mod 7):

$$3^1 \equiv 3,\ \ 3^2 \equiv 2,\ \ 3^3 \equiv 6,\ \ 3^4 \equiv 4,\ \ 3^5 \equiv 5,\ \ 3^6 \equiv 1 \pmod 7$$
$$4^1 \equiv 4,\ \ 4^2 \equiv 2,\ \ 4^3 \equiv 1,\ \ 4^4 \equiv 4,\ \ 4^5 \equiv 2,\ \ 4^6 \equiv 1 \pmod 7,$$

noting, in particular, that $3^6 \equiv 1 \pmod 7$ and $4^6 \equiv 1 \pmod 7$. It turns out, in fact, that if a is *any* integer that's not divisible by 7, then $a^6 \equiv 1 \pmod 7$: for example,

$$2^6 = 64 \equiv 1 \pmod 7 \ \text{ and } \ 10^6 = 1{,}000{,}000 \equiv 1 \pmod 7.$$

These results are special cases of a celebrated result that Pierre de Fermat announced in 1640, but which was first proved by Gottfried Leibniz sometime before 1683. It is usually called 'Fermat's little theorem'.

Fermat's little theorem: If p is a prime, and if a is an integer that is not divisible by p, then $a^{p-1} \equiv 1 \pmod p$.

For example, if $p = 7$ and $a = 17$, then 17 is not divisible by 7, and so $17^6 \equiv 1 \pmod 7$. This is correct, because $17^6 = 24{,}137{,}569$ $= (7 \times 3{,}448{,}224) + 1$.

It can further be shown that if $a^k \equiv 1 \pmod p$, then k must be a factor of $p - 1$.

The idea of the proof of Fermat's little theorem is to look at the first $p - 1$ positive multiples of a and reduce them (mod p). For example, if $p = 7$ and $a = 3$ then the first six positive multiples of a are 3, 6, 9, 12, 15, and 18, and reducing them (mod 7) gives 3, 6, 2, 5, 1, 4. Likewise, if $a = 4$, then the multiples are 4, 8, 12, 16, 20, 24, and reducing them (mod 7) gives 4, 1, 5, 2, 6, 3. In each case we get a rearrangement of the numbers 1, 2, 3, 4, 5, and 6.

In general, the first $p - 1$ positive multiples of a are

$$a, 2a, 3a, 4a, \ldots, (p-1)\,a,$$

and when we reduce them (mod p) the results are all different; this is because if $ma \equiv na \pmod p$, for non-zero numbers m and n, then $m \equiv n \pmod{p}$, after cancelling the term a on both sides (which we can do because a is coprime to p). So these non-zero multiples (mod p) are all different, and must therefore be the numbers $1, 2, 3, \ldots, p-1$, in some order.

We'll now multiply these $p - 1$ multiples together, giving

$$a \times 2a \times 3a \times \ldots \times (p-1)\,a = a^{p-1} \times 1 \times 2 \times 3 \times \ldots \times (p-1).$$

But, as we have just seen, these multiples are just $1, 2, 3, \ldots, p-1 \pmod{p}$ in some order, and their product is

$$1 \times 2 \times 3 \times \ldots \times (p-1) \pmod{p}.$$

Equating these two products gives

$$a^{p-1} \times 1 \times 2 \times 3 \times \ldots \times (p-1) \equiv 1 \times 2 \times 3 \times \ldots \times (p-1) \pmod{p},$$

and cancelling the terms $1, 2, 3, \ldots, p-1$ (which are all coprime to p) gives

$$a^{p-1} \equiv 1 \pmod{p},$$

as we wished to prove.

In the statement of Fermat's little theorem, we made the proviso that a is not divisible by p. We can remove this condition by multiplying both sides of the congruence by a, giving the following alternative form:

> *Fermat's little theorem*: If p is a prime and a is an integer, then $a^p \equiv a \pmod{p}$.

Here we can omit the above proviso, because if a is divisible by p, then the result is still true, because both sides are congruent to $0 \pmod{p}$.

Can we turn things around? Is it true that if $a^n = a \pmod{n}$ for *all* numbers a, then n must be a prime? This is usually the case but, as the American mathematician Robert Carmichael found in 1912, there are exceptions. The smallest example is $n = 561 = 3 \times 11 \times 17$, where $a^{561} \equiv a \pmod{561}$ for all numbers a. Although there are infinitely many such 'Carmichael numbers', they seem to occur quite rarely, with only six others that are less than 10,000, and only forty-three up to one million.

Even if we restrict our attention to $a = 2$, there are exceptions: for example, $2^{341} = 2 \pmod{341}$, but $341 = 11 \times 31$ is not prime. So the ancient Chinese were right in their claim above that n must divide $2^n - 2$ when n is prime, but were wrong to assert that if n divides $2^n - 2$, then n must always be prime.

Counting necklaces

As a brief diversion, we can also derive Fermat's little theorem by counting necklaces! A *necklace* is a circular arrangement of coloured beads. How many different coloured necklaces are there if there are p beads (where p is prime) and a available colours, and if we use at least two colours?

Because of the circular arrangement of beads, each necklace can be counted in p ways, depending on where we start: for example, the necklace in Figure 28 arises from five different strings of beads,

RBRRY, BRRYR, RRYRB, RYRBR, YRBRR,

where R, B, and Y stand for red, blue, and yellow.

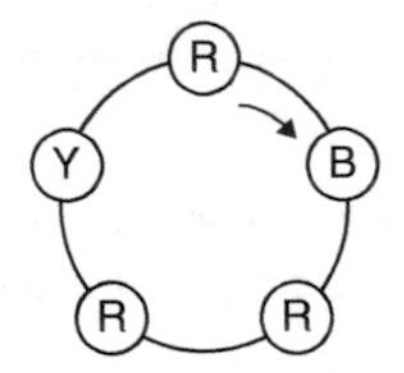

28. A necklace with five beads.

How many different necklaces can there be? Because there are p beads in up to a colours, there are a^p possible strings of beads, or $a^p - a$ strings when we exclude the a single-colour ones (such as

RRRRR) that can each be coloured in only one way. But each necklace arises from p different strings, and so the number of *different* necklaces is $(a^p - a)/p$. Because this must be an integer, $a^p - a$ must be divisible by p—that is, $a^p \equiv a \pmod{p}$, which is Fermat's little theorem.

Shuffling cards

An amusing application of Fermat's result is to the shuffling of cards (sometimes called 'Faro shuffling'). Starting with a normal pack of cards we first cut it into two piles, so that the cards in one pile are in positions 1 to 26 and those in the other pile are in positions 27 to 52. We now shuffle the cards so that the cards in positions 1, 2, 3, . . . , 26 are moved to positions 2, 4, 6, . . . , 52, and the cards in positions 27, 28, 29, . . . , 52 are moved to positions 1, 3, 5, . . . , 51, giving the new ordering

$$27,\ 1,\ 28,\ 2,\ 29,\ 3,\ 30,\ 4, \ldots, 51,\ 25,\ 52,\ 26$$

(see Figure 29). It follows, as we can easily check, that for each a, the card initially in position a has moved to the new position $2a \pmod{53}$.

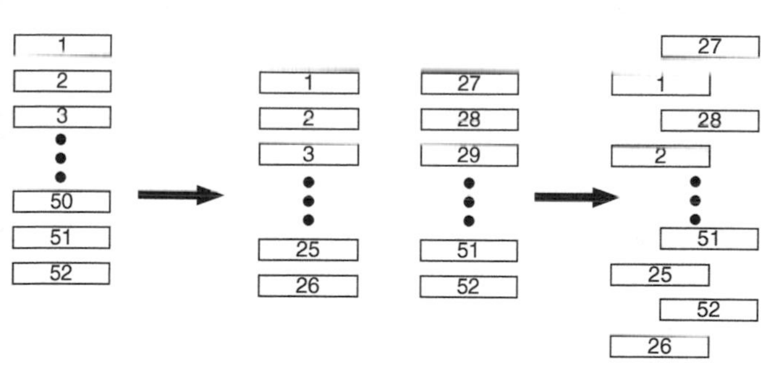

29. Shuffling cards.

How many shuffles are needed to restore the pack of cards to its original order? After n shuffles, the card initially in position a has moved to position $2^n \times a \pmod{53}$, so after the pack has been restored to its original order, $2^n a \equiv a \pmod{53}$ for all numbers a.

Now $\gcd(a, 53) = 1$, so we can cancel the a, giving us $2^n \equiv 1 \pmod{53}$. But by Fermat's little theorem, $2^{52} \equiv 1 \pmod{53}$, so the pack is certainly restored to its original order after 52 shuffles.

But can this happen earlier? If so, the number of shuffles must be a factor of 52, by our remark following the first statement of Fermat's result—that is, 2, 4, 13, or 26. But, after some calculation, we find that

$$2^2 \equiv 4 \pmod{53},\ 2^4 \equiv 16 \pmod{53},\ 2^{13} \equiv 30 \pmod{53},$$
$$2^{26} \equiv 52 \pmod{53}.$$

None of these works, so the order of the cards is restored for the first time after 52 shuffles.

What happens if we now add two Jokers to the pack, making 54 cards in total? Carrying out a similar analysis, we seek the smallest value of n for which $2^n \equiv 1 \pmod{55}$. Now $55 = 5 \times 11$ is not prime, and we can no longer apply Fermat's result directly. But Fermat's little theorem tells us that $2^4 \equiv 1 \pmod{5}$, and so $2^{20} = (2^4)^5 \equiv 1^5 \equiv 1 \pmod{5}$. It also tells us that $2^{10} \equiv 1 \pmod{11}$, and so $2^{20} \equiv 1^2 \equiv 1 \pmod{11}$. Combining these results (which we can, because $\gcd(5, 11) = 1$), we deduce that $2^{20} \equiv 1 \pmod{55}$, and so the order of the cards is certainly restored after 20 shuffles.

Can this happen earlier? If so, the number of shuffles must be a factor of 20—that is, 2, 4, 5, or 10. But

$$2^2 \equiv 4 \pmod{55},\ 2^4 \equiv 16 \pmod{55},\ 2^5 \equiv 32 \pmod{55},$$
$$2^{10} \equiv 34 \pmod{55}.$$

None of these works, so the order of the cards is restored for the first time after 20 shuffles.

Generalizing Fermat's little theorem

We've presented Fermat's little theorem, that if p is a prime number and a is any integer that isn't divisible by p, then $a^{p-1} \equiv 1 \pmod{p}$. But what happens if p is replaced by a composite number? Is Fermat's result still true, and if not, can we adapt it so that it holds in this more general case?

To show that Fermat's result can fail when the modulus is not prime, let's look at congruences (mod 10), and let $a = 3$ (which is coprime to 10). Then $3^{10-1} = 3^9 = 19{,}683$, which is not congruent to 1 (mod 10).

Euler's φ-function

Before we answer the questions above, we'll first need to introduce what Euler called the 'totient function'; the choice of the Greek letter φ (phi) to denote it was made by Gauss in 1801. We define it as follows:

> For each positive integer n, let $\varphi(n)$ be the number of positive integers up to n that are coprime to n—that is, the number of positive integers $a\ (\leq n)$ for which $\gcd(a, n) = 1$.

For example, $\varphi(10) = 4$, because the numbers (up to 10) that are coprime to 10 are 1, 3, 7, and 9; likewise, $\varphi(20) = 8$, where the relevant integers are 1, 3, 7, 9, 11, 13, 17, and 19. A table of values of $\varphi(n)$ for $n \leq 20$ appears in Table 5:

Table 5. Values of $\varphi(n)$, for $n \leq 20$

n	1	2	3	4	5	6	7	8	9	10
$\varphi(n)$	1	1	2	2	4	2	6	4	6	4
n	11	12	13	14	15	16	17	18	19	20
$\varphi(n)$	10	4	12	6	8	8	16	6	18	8

There are a few things to notice here. For a start,

> If p is a prime number, then $\varphi(p) = p - 1$,

because every number up to p is coprime to p, except for p itself.

Also, $\varphi(p^2) = p^2 - p$, because every number up to p^2 is counted, except for the p multiples of p: for example, $\varphi(9) = 3^2 - 3 = 6$, because we count every number up to 9 except for 3, 6, and 9.

Likewise, for any power p^e,

$$\varphi(p^e) = p^e - p^{e-1}.$$

For example,

$$\varphi(16) = \varphi(2^4) = 2^4 - 2^3 = 8.$$

Also, if p and q are different primes, then

$$\varphi(pq) = pq - p - q + 1 = (p-1) \times (q-1) = \varphi(p) \times \varphi(q),$$

because we need to discount the p multiples of q and the q multiples of p, and then restore the number pq (which we had discounted twice). We'll need this result later.

More generally, if $n = ab$, where a and b are coprime, then $\varphi(n) = \varphi(a) \times \varphi(b)$: for example, $\gcd(4, 25) = 1$, and so

$$\begin{aligned}\varphi(100) &= \varphi(4 \times 25) = \varphi(2^2) \times \varphi(5^2) = (2^2 - 2) \times (5^2 - 5) \\ &= 2 \times 20 = 40.\end{aligned}$$

This multiplicative idea holds in general: if n is any product of numbers that are coprime in pairs, then $\varphi(n)$ is the product of the φ-values of the separate numbers: for example,

if $n = 10{,}800 = 2^4 \times 3^3 \times 5^2$, then

$$\begin{aligned}\varphi(n) &= \varphi(2^4) \times \varphi(3^3) \times \varphi(5^2) \\ &= (2^4 - 2^3) \times (3^3 - 3^2) \times (5^2 - 5) \\ &= 8 \times 18 \times 20 = 2880.\end{aligned}$$

We can also rewrite this product as

$$2^3 \times 3^2 \times 5 \times (2-1) \times (3-1) \times (5-1)$$

or as

$$2^4 \times 3^3 \times 5^2 \times (1 - 1/2) \times (1 - 1/3) \times (1 - 1/5),$$

and both of these forms are useful. In general, if the number n is written in canonical form as a product of primes—that is, if

$n = p_1^{e_1} \times p_2^{e_2} \times \ldots \times p_r^{e_r}$, then

$$\varphi(n) = \varphi\left(p_1^{e_1}\right) \times \varphi\left(p_2^{e_2}\right) \times \ldots \times \varphi\left(p_r^{e_r}\right)$$
$$= p_1^{e_1-1} \times p_2^{e_2-1} \times \ldots \times p_r^{e_r-1}$$
$$\times (p_1 - 1) \times (p_2 - 1) \times \ldots \times (p_r - 1).$$

This can also be written as

$$\varphi(n) = p_1^{e_1} \times p_2^{e_2} \times \ldots \times p_r^{e_r}$$
$$\times (1 - 1/p_1) \times (1 - 1/p_2) \times \ldots \times (1 - 1/p_r)$$
$$= n \times (1 - 1/p_1) \times (1 - 1/p_2) \times \ldots \times (1 - 1/p_r):$$

for example, 100 has prime factors 2 and 5, and so

$$\varphi(100) = 100 \times (1 - 1/2) \times (1 - 1/5) = 100 \times 1/2 \times 4/5 = 40,$$

as we saw earlier.

These formulas can be used for calculation, and also to prove some general results about $\varphi(n)$. For example, we can prove that

If $n \geq 3$, then $\varphi(n)$ is even,

as indicated in our table of φ-values. This is because either n is a power of 2 (larger than 2) or it has an odd prime factor p. In the former case, if $n = 2^k$, where $k \geq 2$, then $\varphi(n) = 2^{k-1}$, which is even. In the latter case, the formula for $\varphi(n)$ must include the factor $p - 1$ which is even, and so $\varphi(n)$ is even.

But some even numbers can never appear as φ-values: for example, our table includes the even numbers 2, 4, 6, 8, 10, 12, 16, and 18, but not 14. To see why, suppose that $\varphi(n) = 14$. Now if n has a prime factor p, then $p - 1$ must divide 14, and this can happen only when p is 2 or 3. So n must be 2^r, or 3^s, or $2^r \times 3^s$, for some numbers r and s, and so $\varphi(n) = 2^{r-1}$, or $2 \times 3^{s-1}$, or $2^r \times 3^{s-1}$, respectively. But none of these has 7 as a factor, and so $\varphi(n)$ cannot be 14.

Euler's φ-function has some interesting properties. For example, what do we get if we're given a number n and we add up the φ-values of all its factors? If $n = 10$, then its factors are 1, 2, 5, and 10, and their sum is

$$\varphi(1) + \varphi(2) + \varphi(5) + \varphi(10) = 1 + 1 + 4 + 4 = 10.$$

More generally, it turns out that if we add up the φ-values of the factors of any number n, then we get n itself.

Euler's theorem

We'll now return to the task of extending Fermat's little theorem about the powers of a number (mod p) to powers of a number (mod n), where n may be composite. When we introduced Fermat's result we started by looking at the powers of certain numbers (mod 7). We'll now imitate the process by looking at the first few powers of certain numbers (mod 9):

$$\begin{array}{llllll} 2^1 \equiv 2, & 2^2 \equiv 4, & 2^3 \equiv 8, & 2^4 \equiv 7, & 2^5 \equiv 5, & 2^6 \equiv 1 \pmod 9, \\ 3^1 \equiv 3, & 3^2 \equiv 0, & 3^3 \equiv 0, & 3^4 \equiv 0, & 3^5 \equiv 0, & 3^6 \equiv 0 \pmod 9, \\ 4^1 \equiv 4, & 4^2 \equiv 7, & 4^3 \equiv 1, & 4^4 \equiv 4, & 4^5 \equiv 7, & 4^6 \equiv 1 \pmod 9, \\ 7^1 \equiv 7, & 7^2 \equiv 4, & 7^3 \equiv 1, & 7^4 \equiv 7, & 7^5 \equiv 4, & 7^6 \equiv 1 \pmod 9. \end{array}$$

It turns out that, apart from the powers of 0, 3, and 6, every 6th power is 1.

Now $6 = \varphi(9)$, and we can summarize this by saying:

If a is coprime to 9, then $a^{\varphi(9)} \equiv 1 \pmod 9$.

This result is a special case of the result we were seeking:

Euler's theorem: If n is a positive integer, and if a is any integer with $\gcd(a, n) = 1$, then $a^{\varphi(n)} \equiv 1 \pmod n$.

For example, if $n = 20$ and $a = 13$, then $\varphi(20) = 8$ and $\gcd(13, 20) = 1$, and we deduce that $13^8 \equiv 1 \pmod{20}$, which is correct because

$$13^8 = 815{,}730{,}721 = (20 \times 40{,}786{,}536) + 1.$$

To see why Euler's theorem is true, let's imitate what we did earlier and look at the first few positive multiples of the numbers coprime to 9—that is, 1, 2, 4, 5, 7, and 8—and reduce them (mod 9). Multiplying these numbers by 4 gives 4, 8, 16, 20, 28, 32, and reducing them (mod 9) gives 4, 8, 7, 2, 1, 5. Likewise, multiplying them by 7 gives 7, 14, 28, 35, 49, 56, and reducing them (mod 9) gives 7, 5, 1, 8, 4, 2. In each case we get a rearrangement of the numbers 1, 2, 4, 5, 7, and 8.

To prove Euler's theorem in this case, we'll mimic the proof of Fermat's little theorem. Multiplying the numbers 1, 2, 4, 5, 7, and 8 by a (where $\gcd(a, 9) = 1$) gives $1a$, $2a$, $4a$, $5a$, $7a$, and $8a$, and reducing these (mod 9) gives 1, 2, 4, 5, 7, and 8, though possibly in a different order. So, on multiplying them all together, we get

$$1a \times 2a \times 4a \times 5a \times 7a \times 8a = 1 \times 2 \times 4 \times 5 \times 7 \times 8 \pmod 9.$$

Cancelling the numbers 1, 2, 4, 5, 7, and 8 (which are all coprime to 9) then gives $a^6 = 1 \pmod 9$, as we saw above.

The general proof is similar. We begin by listing the $\varphi(n)$ numbers up to n that are coprime to n—we'll call them $b_1, b_2, \ldots, b_k$, where $k = \varphi(n)$. We then multiply them by a, to give the multiples $b_1 a$, $b_2 a, \ldots, b_k a$, and reduce them (mod n). The results will all be different, and must therefore be $b_1, b_2, \ldots, b_k$ in some order.

Multiplying these multiples of a together and equating the two products, as above, gives

$$a^{\varphi(n)} \times b_1 \times b_2 \times \ldots \times b_k \equiv b_1 \times b_2 \times \ldots \times b_k \pmod n.$$

Cancelling the terms $b_1, b_2, \ldots, b_k$ (which are all coprime to n) then gives

$$a^{\varphi(n)} \equiv 1 \pmod n,$$

as we wished to prove.

We conclude this section by noting that if n is a prime number p, then $\varphi(n) = p - 1$, and we get $a^{p-1} \equiv 1 \pmod{p}$, which is Fermat's little theorem, as we'd expect.

Factorizing large numbers

Before we see how Euler's theorem makes its appearance in cryptography, let's consider briefly the subject of trying to factorize a number into its prime factors.

Many real-life processes are irreversible: for example, it's easy to squeeze toothpaste from a tube but difficult to reverse the operation, and it's easy to break an egg but impossible to 'unbreak' it, especially if it's Humpty Dumpty.

Another example is the factorization of large numbers. It's generally straightforward to multiply two primes together, but if we're given the product, then it's often more difficult to factorize it into its two prime constituents: for example, we can easily multiply 23 and 89 to give 2047, but given the number 2047 it may take us some time to find its factors by hand. If the primes are large—say, 250 digits—then their product can easily be calculated by machine, but no current computer can factorize their 500-digit product in a reasonable amount of time.

A traditional method for testing whether a given number n is prime is due to Fermat, and works particularly well when n is the product of two numbers that are close to each other. The method is based on the fact that

$$\text{if } n = a^2 - b^2 = (a+b)(a-b), \text{then } a^2 - n = b^2.$$

To apply Fermat's method, we first find the smallest integer m that's $\sqrt{n}$ or larger, and we calculate the numbers

$$m^2 - n,\ (m+1)^2 - n,\ (m+2)^2 - n,\ \text{and so on,}$$

until we reach a square. Then, if $(m+k)^2 - n = b^2$ (say), we have

$$n = (m+k)^2 - b^2 = (m+k+b)\times(m+k-b),$$

which gives a factorization.

For example, if $n = 6077$, we check that $\sqrt{6077} = 77.995\ldots$, and so we take $m = 78$. Then:

$$78^2 - 6077 = 7 \quad \text{(not a square)}$$
$$79^2 - 6077 = 164 \quad \text{(not a square)}$$
$$80^2 - 6077 = 323 \quad \text{(not a square)}$$
$$81^2 - 6077 = 484 \quad \text{(the square of 22)}.$$

So $6077 = 81^2 - 22^2 = (81+22)\times(81-22) = 103\times 59$.

Fermat used his method to factorize the number $n = 2{,}027{,}651{,}281$.

Noting that $\sqrt{n} = 45{,}029.449\ldots$, he took $m = 45{,}030$ and calculated as follows:

$$45{,}030^2 - n = 49{,}619, \quad 45{,}031^2 - n = 139{,}680,$$
$$45{,}032^2 - n = 229{,}743, \quad 45{,}033^2 - n = 319{,}808,$$
$$45{,}034^2 - n = 409{,}875 \quad 45{,}035^2 - n = 499{,}944,$$
$$45{,}036^2 - n = 590{,}015, \quad 45{,}037^2 - n = 680{,}088,$$
$$45{,}038^2 - n = 770{,}163, \quad 45{,}039^2 - n = 860{,}240,$$
$$45{,}040^2 - n = 950{,}319, \quad 45{,}041^2 - n = 1{,}040{,}400.$$

Now $1{,}040{,}400 = 1020^2$, and so

$$n = 45{,}041^2 - 1020^2 = (45{,}041 + 1020)\times(45{,}041 - 1020)$$
$$= 46{,}061 \times 44{,}021,$$

giving him a factorization.

Fermat's approach to factorization is the basis of several rather more sophisticated methods for factorizing large numbers. One of these is called the *quadratic sieve method* which seeks some numbers among the above differences whose product is a square, rather than simply being a square. Another method, due to Euler and based on an idea of Mersenne, led to the factorization $1{,}000{,}009 = 3413 \times 293$. Many other methods have been described, but no efficient algorithm has ever been discovered.

RSA public key cryptography

We've just seen that multiplying two prime numbers is comparatively simple, but factorizing a large number into its prime factors can be extremely difficult. It's from this asymmetric process that a method for secretly encrypting messages has been devised. It was proposed in 1973 by Clifford Cocks, formerly of the secret World War II codebreaking unit at Bletchley Park in England, but remained classified until 1997. Independently rediscovered in 1978 by Ron Rivest, Adi Shamir, and Leonard Adleman, it's now known by their initials as *RSA encryption*. It's essentially this method that is used to preserve intelligence information, and it provides us with an excellent example of how results in pure mathematics that were investigated for their own interest (such as Euler's theorem) can later be applied in unexpected ways in highly practical situations.

Suppose that Alice wishes to send a very secret message to Bob, in such a way that no eavesdropper who might intercept it can decode it. Bob first selects two large primes, p and q, and calculates their product, $N = pq$. He also chooses a number e that is coprime to $\varphi(N)$—that is,

$$\gcd(e, (p-1)(q-1)) = 1.$$

Bob then publicly announces the numbers e and N, but he doesn't release N's factors, p and q. The numbers e and N are the *public key* that's known to all, while Bob remains the only person who knows p and q, and therefore $\varphi(N)$.

Alice can now send her secret message. She first converts it to a numerical form—for example, by writing $A = 00, B = 01, \ldots,$ $Z = 25$—and calls the resulting message M. Knowing the numbers e and N, she can then calculate the number $E \equiv M^e \pmod{N}$, and send it to Bob, the only person who can decrypt it. But how can he do so?

To recover M, Bob first calculates a number m for which $me \equiv 1 \pmod{\varphi(N)}$. To do so, he can use Euclid's algorithm from Chapter 2: because $\gcd(e, \varphi(N)) = 1$, he can find integers m and n for which $1 = (m \times e) + (n \times \varphi(N))$, and so $me \equiv 1 \pmod{\varphi(N)}$.

Then, on receiving the encrypted message E and knowing m, he calculates

$$E^m \equiv (M^e)^m \equiv M^{me} \equiv M^{1-n\varphi(N)} \equiv M \times M^{-n\varphi(N)} \pmod{N}.$$

But, by Euler's theorem, $M^{\varphi(N)} \equiv 1 \pmod{N}$, and so $M^{-n\varphi(N)} \equiv 1^{-n} \equiv 1 \pmod{N}$. It follows that $E^m \equiv M \pmod{N}$, and so all that Bob needs to do is to calculate $E^m \pmod{N}$ in order to retrieve Alice's original message M.

As an example of the calculations involved, suppose that the public key consists of the numbers $e = 11$ and $N = 1073$. Bob knows that $1073 = 29 \times 37$, and so he can calculate

$$\varphi(N) = \varphi(29) \times \varphi(37) = 28 \times 36 = 1008,$$

and check that $\gcd(e, \varphi(N)) = \gcd(11, 1008) = 1$, as required.

The congruence $me \equiv 1 \pmod{\varphi(N)}$ now becomes $11m \equiv 1 \pmod{1008}$. By using Euclid's algorithm, Bob finds that

$$1 = (275 \times 11) - (3 \times 1008),$$

so that $275 \times 11 \equiv 1 \pmod{1008}$, and so he takes $m = 275$. Having received Alice's coded message as E, Bob now calculates $E^{275} \pmod{1073}$, and so recaptures her original message.

Chapter 7
Conjectures and theorems

In this chapter we'll explore some further topics that are related to prime numbers. We'll start with two problems that we met in Chapter 1, on sums of primes and twin primes, and follow this by investigating the distribution of prime numbers. We then turn our attention to lists of primes that are equally spaced, and to a fuller exploration of unique factorization. This chapter presents several of the deepest results in number theory, and includes some examples of recent work in the subject.

Two famous conjectures

In this section we explore two conjectures on primes that are very easy to state but have never yet been proved. Their difficulty is linked to the fact that they involve addition or subtraction, whereas the primes are mainly concerned with multiplication.

Goldbach's conjecture

On 7 June 1742, the German mathematician Christian Goldbach wrote a letter to Euler about writing numbers as the sum of primes. This contained a claim that is now known as 'Goldbach's conjecture', but which Euler described as 'a completely certain theorem, although I cannot prove it':

Goldbach's conjecture: Every even number that is larger than 2 can be written as the sum of two primes.

In Chapter 1 you saw some instances of this, and further examples are $100 = 97 + 3$ and $1000 = 509 + 491$.

Goldbach's conjecture is now known to be true for all even numbers up to 4×10^{18}, and so finding a counter-example seems extremely unlikely.

A major advance in settling Goldbach's conjecture was made in 1966 by the Chinese mathematician Chen Jingrun, who proved that every even number that is larger than 2 can be written as the sum of a prime number and an 'almost prime'—that is, another number that is either a prime or a number with just two prime factors. His work involved some systematic sieve methods, an area of number theory whose origins can be traced back to the sieve of Eratosthenes which we met in Chapter 3.

A result that seems related to Goldbach's conjecture had already been proved a few years earlier, in 1937, by the Russian mathematician Ivan Matveevich Vinogradov:

From some point onwards, every odd number can be written as the sum of three primes.

Here, the phrase 'from some point onwards' might seem to indicate that only a small number of cases remain to be checked—which is a finite task that could quickly be carried out by hand, or by computer. But in practice, the point from which the result had been proved true was massive, with millions of digits, and checking the remaining cases was way beyond the capacity of any existing computers. However, in 2013, and following much further theoretical work by several people, the Peruvian mathematician Harald Helfgott (working in France, with computational help from the British number theorist Dave Platt) managed to reduce, and eventually to eliminate, the huge gap between what had been proved for large numbers and what was already known for small numbers, giving the following result:

> Every odd number that is larger than 5 can be written as the sum of three primes.

What has this to do with Goldbach's conjecture? Well, if Goldbach's conjecture is true, then every even number (≥ 4) is the sum of at most two primes p and q, and adding 3 tells us that every odd number (≥ 7) is the sum of at most three primes (p, q, and 3). But unfortunately the argument doesn't work the other way around: Helfgott's achievement doesn't yield a proof of Goldbach's conjecture, but it does provide a weaker form of it. For if n is an even number that is larger than 8, then $n - 3$ can be written as the sum of three primes. Also, $8 = 2 + 2 + 2 + 2$, and so

> Every even number that is larger than 6 can be written as the sum of four primes.

The twin prime conjecture

The second conjecture concerns twin primes which, as you saw in Chapter 1, are pairs of primes that differ by 2. Those up to 100 are

> 3 and 5, 5 and 7, 11 and 13, 17 and 19, 29 and 31, 41 and 43, 59 and 61, 71 and 73.

There are thirty-five pairs of twin primes up to one thousand, over eight thousand pairs up to one million, and over three million pairs up to one billion. The largest known pair has over fifty thousand digits!

In 1846 the following conjecture was formulated by the French number theorist Alphonse de Polignac:

> *Twin prime conjecture*: There are infinitely many pairs of twin primes.

For many years, number theorists have tried to prove this, but with little success. Then, in 1966, and associated with his work on

Goldbach's conjecture, Chen Jingrun used sieve methods to prove that there are infinitely many prime numbers p for which $p + 2$ is either a prime or an almost prime.

Among other more recent investigations were proofs that the twin prime conjecture is true if one is allowed to assume certain additional results. Then, suddenly and unexpectedly, a major breakthrough was made in June 2013 by the Chinese-born American mathematician Yitang Zhang. Using some of this earlier work, but without needing to assume any other results, he proved that infinitely many pairs of prime numbers differ by at most 70 million. This is a very far cry from the desired result, with 70 million instead of 2, but it was the first result of its kind and it created a whole cottage industry of mathematicians seeking to reduce the gap to something more manageable.

At first, the gap came down from 70 million to around 42 million, and then something rather remarkable happened. Mathematicians are used to writing up their research and then publishing it in polished form in journals, a process that can take a year or more: this means that much time may elapse before their results become widely known. But from around 2009 various mathematicians—notably, Tim Gowers from Cambridge and Terry Tao from Los Angeles—had proposed a more collaborative approach, known as the *Polymath project*, in which contributors from around the world could work on problems publicly by pooling their ideas, feeding in comments, and suggesting improvements. Advances could thereby be made and shared at a much faster rate, while individual contributors would still receive due credit for their ideas.

In June 2013 Tao initiated *Polymath8*, a project entitled 'Bounded gaps between primes', in which contributors were invited to improve Zhang's result. Within a few weeks, with contributions from many people, the gap had decreased from 42 million to

387,620, and then to about 12,000, and after another month to 4680.

This was followed by a lull in activity, and new ideas were needed. By this time, James Maynard, who had gained his doctorate in Oxford and was then in Montreal, appeared on the scene with a different approach, also discovered independently by Tao. By the end of 2013 Maynard had reduced the gap to 600 and, as he wrote at the time:

> I was surprised at how much time I ended up devoting to the Polymath project. This was partly because the nature of the project was so compelling—there were clear numerical metrics of 'progress' and always several possible ways of obtaining small improvement, which was continually encouraging. The general enthusiasm amongst the participants (and others outside of the project) also encouraged me to get more and more involved in the project.

At the time of writing, the current gap is 246. This is a massive improvement on the original 70 million—but there's still quite a way to go before the twin prime conjecture is finally laid to rest.

The distribution of primes

How are the prime numbers distributed? Although they generally seem to 'thin out', the further along the list we proceed, they don't seem to be distributed in a regular manner. For example, the hundred numbers just below 10 million include nine primes,

$$9{,}999{,}901, \quad 9{,}999{,}907, \quad 9{,}999{,}929, \quad 9{,}999{,}931,$$
$$9{,}999{,}937, \quad 9{,}999{,}943, \quad 9{,}999{,}971, \quad 9{,}999{,}973,$$
$$9{,}999{,}991,$$

whereas the hundred numbers just above it include only two,

$$10,000,019 \quad \text{and} \quad 10,000,079.$$

But although twin primes seem to crop up however far we go, we can also construct arbitrarily long strings of numbers that aren't prime. To do so, we'll use the *factorial numbers* $n!$, defined by

$$1! = 1, \;\; 2! = 2 \times 1 = 2, \;\; 3! = 3 \times 2 \times 1 = 6,$$
$$4! = 4 \times 3 \times 2 \times 1 = 24,$$

and in general,

$$n! = n \times (n-1) \times (n-2) \times \ldots \times 3 \times 2 \times 1.$$

Now, because the number $n!$ is divisible by all the numbers from 1 to n, we see that:

$$2 \text{ divides } n! + 2, \;\; 3 \text{ divides } n! + 3, \;\; 4 \text{ divides } n!, \ldots,$$
$$\text{and } \; n \text{ divides } n! + n,$$

and so the numbers

$$n! + 2, \;\; n! + 3, \;\; n! + 4, \;\; \ldots, \;\; n! + (n-1), \;\; n! + n$$

are all composite, giving us a string of $n - 1$ composite numbers. Another example is:

$$n! - n, \;\; n! - (n-1), \;\; \ldots, \;\; n! - 3, \;\; n! - 2.$$

So, for example, two strings of 1000 successive composite numbers are

$$1001! + 2, \;\; 1001! + 3, \;\; \ldots, \;\; 1001! + 1000, \;\; 1001! + 1001$$

and

$$1001! - 1001, \;\; 1001! - 1000, \ldots, 1001! - 3, \;\; 1001! - 2.$$

The prime number theorem

In his inaugural lecture as professor at the University of Bonn in 1975, the distinguished number-theorist Don Zagier remarked:

> There are two facts of which I hope to convince you so overwhelmingly that they will permanently be engraved in your hearts.
>
> The first is that the prime numbers belong to the most arbitrary objects studied by mathematicians: they grow like weeds, seeming to obey no other law than that of chance, and nobody can predict where the next one will sprout.
>
> The second fact is even more astonishing, for it states just the opposite: that the prime numbers exhibit stunning regularity, that there are laws governing their behaviour, and that they obey these laws with almost military precision.

To see what he meant by this, we'll introduce the prime-counting function $\pi(x)$, which counts the number of primes up to any number x. (This use of the Greek letter π (pi) is nothing to do with the circle number π.) So $\pi(10) = 4$, because there are exactly four primes (2, 3, 5, and 7) up to 10, and $\pi(20) = 8$, because there are four more primes (11, 13, 17, and 19). Continuing, we find that $\pi(100) = 25$, $\pi(1000) = 168$, and $\pi(10,000) = 1229$.

If we plot the values of the primes up to 100 on a graph we get a jagged pattern—each new prime creates a jump (see Figure 30). But if we stand back and view the primes up to 100,000, we get a lovely smooth curve—the primes do indeed seem to increase very regularly.

We can describe this overall regularity more precisely by comparing the values of x and $\pi(x)$ as x increases. We get the following Table 6 lists x, $\pi(x)$, and their ratio $x/\pi(x)$ (to one decimal place).

So up to 100 one-quarter of the numbers are prime, up to 1000 about one-sixth of them are prime, and so on. We can express this 'thinning-out' more precisely by noting that whenever x is multiplied by 10, the ratio $x/\pi(x)$ seems to increase by around 2.3.

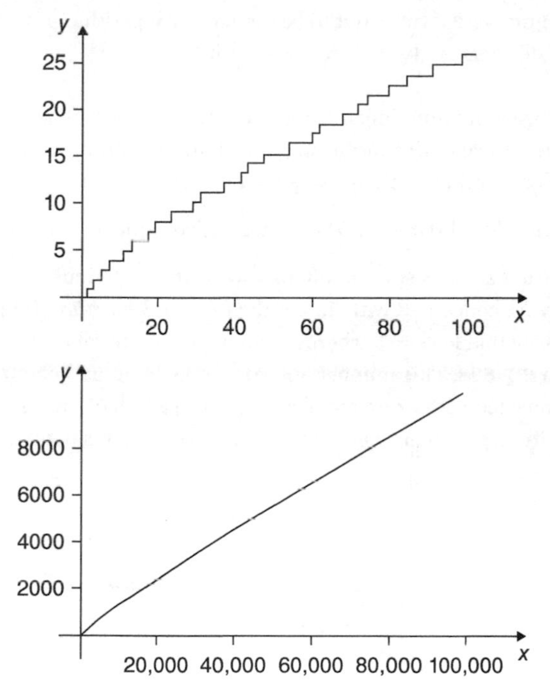

30. The distribution of primes.

Table 6. x, $\pi(x)$, and $x/\pi(x)$, when x is a power of 10

x	$\pi(x)$	$x/\pi(x)$
10	4	2.5
100	25	4.0
1000	168	6.0
10,000	1229	8.1
100,000	9592	10.4
1,000,000	78,498	12.7
10,000,000	664,579	15.0
100,000,000	5,761,455	17.4
...	...	...

This number 2.3 turns out to be the natural logarithm of 10. So what do we mean by 'natural logarithm'?

The logarithm function, introduced in the early 1600s, is a mathematical device for turning multiplication problems into simpler addition ones, by using the basic rule:

$$\log(a \times b) = \log a + \log b, \text{for any positive numbers } a \text{ and } b,$$

There are actually several different logarithm functions, but the one we're concerned with here is the *natural logarithm*. It has the property that $\log e = 1$, where e is an important number that's about 2.71828. This number appears throughout mathematics and is connected with exponential growth. The graph of the natural logarithm, $y = \log x$ (sometimes written as $\ln x$), is shown in Figure 31.

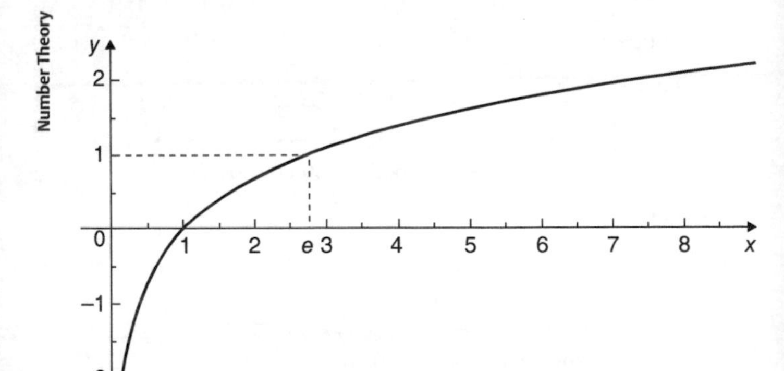

31. The graph of the natural logarithm.

Because the natural logarithm x turns multiplication into addition, and, in particular,

$$\log\ (10x) = \log x + \log 10 = \log x + 2.3\,(\text{approximately}),$$

we can explain the phenomenon illustrated in the above table of values by saying that, as x increases, $\pi(x)$ behaves rather like $x/\log x$—or, more precisely, that their ratio approaches 1 as x becomes large. Figure 32 shows the similarity between the graphs of $\pi(x)$ and $x/\log x$.

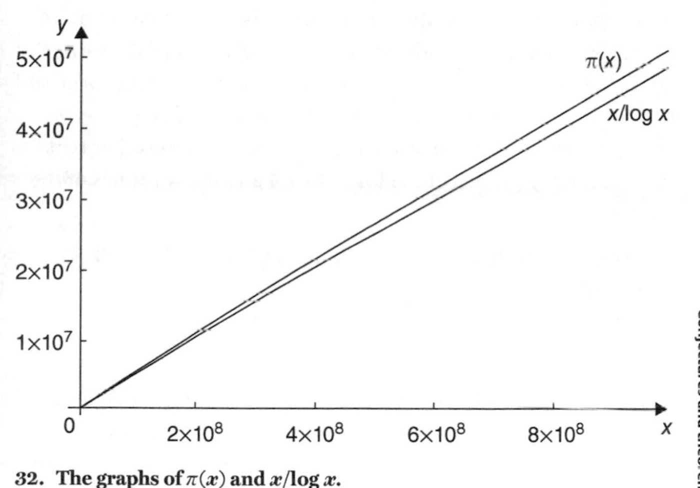

32. The graphs of $\pi(x)$ and $x/\log x$.

Gauss guessed this connection (and even closer approximations) while experimenting with prime numbers at the age of 15. But it wasn't proved until 1896, when Jacques Hadamard (from France) and Charles de la Vallée Poussin (from Belgium) did so independently, using sophisticated ideas from an area of calculus called complex analysis. It is known as the 'prime number theorem'.

> *Prime number theorem*: As x increases indefinitely, the ratio $\pi(x)/(x/\log x)$ tends to 1.

It then took a further fifty years until the Norwegian Atle Selberg and the Hungarian Paul Erdős discovered a purely number-theoretical proof—and because Hadamard lived to 98, de la Vallée

Poussin lived to 96, and Selberg lived to 90, it seems that proving the prime number theorem assists longevity!

Primes in arithmetic progressions

In Chapter 3 you met Euclid's proof that there are infinitely many prime numbers, and we shall now adapt his ideas to derive more precise information. Recalling that every number has the form $4q, 4q + 1, 4q + 2$, or $4q + 3$, we note that numbers of the form $4q$ and those of the form $4q + 2$ (other than 2) cannot be prime, because they're even and so have 2 as a factor. On the other hand, there are many primes of the form $4q + 1$ and many primes of the form $4q + 3$:

primes of the form $4q + 1$: 5, 13, 17, 29, 37, 41, 53, 61, 73, 89, 97, 101, ...
primes of the form $4q + 3$: 3, 7, 11, 19, 23, 31, 43, 47, 59, 67, 71, 79, 83, 103, ...

By modifying Euclid's argument slightly we can prove that there are infinitely many prime numbers of the form $4q + 3$. To do so, we'll assume (for a contradiction) that there are only finitely many primes, $p_1, p_2, \ldots, p_n$, of this form (other than 3), but this time we'll consider the number

$$N = 4 \times (p_1 \times p_2 \times \ldots \times p_n) + 3,$$

which certainly has the form $4q + 3$. Now, each of our primes divides the product $p_1 \times p_2 \times \ldots \times p_n$, and so cannot also divide N. So either N is a new prime of the form $4q + 3$, or it's a composite number, in which case it must split into new primes. But these new primes cannot all be of the form $4q + 1$, because multiplying any numbers of that form always gives another one:

$$\begin{aligned}(4k + 1) \times (4l + 1) &= 16kl + 4k + 4l + 1 \\ &= 4 \times (4kl + k + l) + 1.\end{aligned}$$

So at least one of the new primes must have the form $4q + 3$. It follows in either case that there's a prime of the form $4q + 3$ other than $p_1, p_2, \ldots$, and p_n, and this contradiction shows that there must be infinitely many prime numbers of this form.

We can imitate this proof to show that there are infinitely many prime numbers of the form $6q + 5$. We assume that there are only finitely many primes, $p_1, p_2, \ldots, p_n$, of this form, but this time we consider the number

$$N = 6 \times (p_1 \times p_2 \times \ldots \times p_n) + 5,$$

noting that multiplying any numbers of the form $6q + 1$ always gives another one. We deduce that there are infinitely many primes of the form $6q + 5$.

We can also adapt Euclid's proof to deduce that there are infinitely many primes of certain other forms. But we cannot use this approach to prove that there are infinitely many primes of the form $4q + 1$, because such numbers can also be obtained by multiplying only numbers of the form $4q + 3$: for example, $21 = 3 \times 7$. We're likewise unable to adapt this approach to prove that there are infinitely many primes of the form $6q + 1$, because such numbers can be obtained by multiplying only numbers of the form $6q + 5$: for example, $55 = 5 \times 11$.

But there are indeed infinitely many prime numbers of the form $4q + 1$, and we can prove this by recalling from Chapter 4 that if -1 is a square (mod p), then p must be a prime of the form $4q + 1$. To get a contradiction we'll assume that there are finitely many primes, $p_1, p_2, \ldots, p_n$, of the form $4q + 1$, and this time we'll consider the number

$$N = 4 \times (p_1 \times p_2 \times \ldots \times p_n)^2 + 1,$$

which certainly has the form $4q + 1$. As before, none of our primes $p_1, p_2, \ldots, p_n$ can divide N, so N is either a new prime or is a product of new primes. In either case there's a new prime, which we'll call p, that divides N. It follows that

$$4 \times (p_1 \times p_2 \times \ldots \times p_n)^2 + 1 \equiv 0 \pmod{p},$$
$$4 \times (p_1 \times p_2 \times \ldots \times p_n)^2 \equiv -1 \pmod{p}.$$

-1 is a square (mod p) and we deduce that p has the form so $4q + 1$. This contradiction proves there must be infinitely many primes of the form $4q + 1$. A similar proof shows that that there are infinitely many primes of the form $6q + 1$.

Having shown that there are infinitely many primes of these forms, we may now ask whether there are infinitely many prime numbers of the form $aq + b$, for any given numbers a and b. In its most general form, the answer to this question is 'no', for if a and b have a common factor d that's greater than 1, then d must also divide all numbers of the form $aq + b$. For example, there are no primes of the form $6q + 4$ (because all such numbers have 2 as a factor) or of the form $9q + 6$ (because all such numbers have 3 as a factor).

But if a and b are coprime, then we have the following result, conjectured in 1785 by Legendre and proved in 1837 by Lejeune Dirichlet.

> *Dirichlet's theorem*: If a and b are given numbers with $\gcd(a, b) = 1$, then there are infinitely many prime numbers of the form $aq + b$, where q is an integer.

For example, there are infinitely many prime numbers of the form $4q + 1$, because $\gcd(4, 1) = 1$, and there are infinitely many prime numbers of the form $89q + 55$, because $\gcd(89, 55) = 1$. Also, because $\gcd(10, 9) = 1$, there are infinitely many prime numbers of the form $10q + 9$—that is, there are infinitely many primes with final digit 9—and likewise there are infinitely many primes with final digit 1, 3, or 7.

An *arithmetic progression* with first term b and common difference k is a finite or infinite sequence of equally spaced numbers of the form

$$b, b + k, b + 2k, b + 3k, \ldots.$$

For example, the sequence of numbers

$$5, \quad 11, \quad 17, \quad 23, \quad 29$$

is a finite arithmetic progression with first term $b = 5$ and common difference $k = 6$, and the sequence

$$3, \quad 7, \quad 11, \quad 15, \quad 19, \quad 23, \ldots$$

of numbers of the form $4q + 3$ is an infinite arithmetic progression with first term $b = 3$ and common difference $k = 4$. More generally, the sequence of numbers of the form $aq + b$ is an infinite arithmetic progression with first term b and common difference a, and if $\gcd(a, b) = 1$, then such a sequence must include infinitely many primes, by Dirichlet's theorem.

Let's now turn the above question around. The sequence

$$5, \quad 11, \quad 17, \quad 23, \quad 29$$

is an arithmetic progression of five primes with first term 5 and common difference 6, but it cannot be extended to an arithmetic progression of six primes because the next term is the composite number $35 = 5 \times 7$. An example of an arithmetic progression of six primes is

$$7, \quad 37, \quad 67, \quad 97, \quad 127, \quad 157.$$

More generally, we can ask:

Does the list of primes contain arithmetic progressions of any chosen length n?

As we've just seen, the answer to this is 'yes' when $n = 5$ and 6. For $n = 10$, the following arithmetic progression with first term 199 and common difference 210 consists entirely of primes:

$$199, \ 409, \ 619, \ 829, \ 1039, \ 1249, \ 1459, \ 1669, \ 1879, \ 2089.$$

Likewise, for $n = 23$, the arithmetic progression with first term 56,211,383,760,397 and common difference 44,546,738,095,860 consists entirely of primes. At the time of writing, the longest known arithmetic progressions of primes contain 26 primes.

The general question was answered in the affirmative by Ben Green and Terry Tao in 2004:

> *The Green–Tao theorem*: Given any number n, the list of prime numbers contains an arithmetic progression of n primes.

Unique factorization

In Chapter 3 we saw that the factorization of positive integers into primes is unique, apart from the order of the factors—it's a basic property of our number system. But it doesn't hold for certain other systems of numbers. Here are two examples:

Example 1. Consider just the positive even numbers (which we'll call *e-numbers*),

$$2, 4, 6, 8, 10, 12, 14, 16, \ldots,$$

and consider the factorization of *e*-numbers into smaller e-numbers. We'll call an e-number *e-composite* if it can be written as a product of smaller e-numbers, and *e-prime* if not. So 16 and 24 are e-composite numbers because $16 = 2 \times 8$ and $24 = 4 \times 6$, but 6 and 10 are e-prime numbers because they can't be written as products of smaller e-numbers. The first few e-primes are

$$2, 6, 10, 14, 18, 22, 26, 30, 34, 38, 42, \ldots .$$

But in this number system we don't have unique factorization into e-primes: for example, 2, 6, and 18 are all e-primes, and the *e*-number 36 can be written as either 6×6 or 2×18.

Example 2 (due to David Hilbert). Consider just the numbers of the form $4k + 1$ (*h-numbers*),

$$5, 9, 13, 17, 21, 25, 29, 33, \ldots,$$

and consider the factorization of h-numbers into smaller h-numbers. We'll call an h-number *h-composite* if it can be written as a product of smaller h-numbers, and *h-prime* if not. So 25 and 45 are h-composite numbers because $25 = 5 \times 5$ and $45 = 5 \times 9$, but 9 and 21 are h-prime numbers because they can't be written as products of smaller h-numbers. The first few h-primes are

$$5, 9, 13, 17, 21, 29, 33, 37, 41, 49, \ldots .$$

But in this number system we don't have unique factorization into h-primes: for example, 9, 21, and 49 are all h-primes, and the h-number 441 can be written as either 9×49 or 21×21.

However, there are some number systems, other than the positive integers, where we do have unique factorization. We give two examples:

Example 3. Consider the numbers of the form $a + b\sqrt{2}$, where a and b are integers ($\sqrt{2}$-*numbers*). Such numbers include $2 + 3\sqrt{2}$ and $1 + 4\sqrt{2}$, and we can carry out ordinary arithmetic with them, replacing $(\sqrt{2})^2$ wherever it arises by 2:

$$\begin{aligned} \text{addition:} \quad & (2 + 3\sqrt{2}) + (1 + 4\sqrt{2}) \\ & = (2 + 1) + (3 + 4)\sqrt{2} = 3 + 7\sqrt{2}, \\ \text{multiplication:} \quad & (2 + 3\sqrt{2}) \times (1 + 4\sqrt{2}) \\ & = 2 + (8 + 3)\sqrt{2} + 12(\sqrt{2})^2 = 26 + 11\sqrt{2}. \end{aligned}$$

In this system we can define $\sqrt{2}$-prime and $\sqrt{2}$-composite numbers: for example, $26 + 11\sqrt{2}$ is $\sqrt{2}$-composite, because it can be written as $\left(2 + 3\sqrt{2}\right) \times \left(1 + 4\sqrt{2}\right)$. It is less easy to decide which are the $\sqrt{2}$-primes, but this can be done, and we can prove that every $\sqrt{2}$-number can be written as a product of $\sqrt{2}$-primes in only one way, apart from the order in which they appear.

Example 4. This is similar to the previous example, except that we replace $\sqrt{2}$ by i, the (imaginary) square root of -1. These numbers of the form $a + bi$, where a and b are integers, were introduced by

Gauss in 1801, and are known as *Gaussian integers*. Such numbers include $2 + 3i$ and $1 + 4i$, and we can carry out arithmetic with them, replacing i^2 wherever it arises by -1: for example,

$$\begin{aligned} \text{addition:} \quad & (2+3i)+(1+4i) \\ & = (2+1)+(3+4)i = 3+7i, \\ \text{multiplication:} \quad & (2+3i)\times(1+4i) \\ & = 2+(8+3)i+12i^2 = -10+11i. \end{aligned}$$

In this system Gauss defined Gaussian primes and Gaussian composite numbers, and proved that every Gaussian integer can be written as a product of Gaussian primes in just one way, apart from the order in which the Gaussian primes appear.

We can imitate the ideas of Examples 3 and 4 to explore prime and composite numbers of the form $a + b\sqrt{n}$, where n is an integer. We'll assume that n is 'square-free'—that is, it has no square factors other than 1, because we can simply remove them: for example, because $18 = 2 \times 3^2$ we can replace $\sqrt{18}$ by $\sqrt{2}$.

We sometimes have unique factorization into primes, as happened for $n = 2$ and $n = -1$, but not always. When n is positive, it is not known in general when there's factorization into primes in just one way. But when n is negative, we can give a complete answer. As before we don't always get unique factorization: for example, when $n = -5$, the numbers, $2, 3, 1 + \sqrt{-5}$, and $1 - \sqrt{-5}$ all play the role of primes, and yet we can write

$$6 = 2 \times 3 = \left(1 + \sqrt{-5}\right) \times \left(1 - \sqrt{-5}\right),$$

so the factorization into primes is not unique in this case.

In his *Disquisitiones Arithmeticae* Gauss showed that there is unique factorization when $n = -1$ (the Gaussian integers), and for a few other negative square-free values which he listed. He believed these to be the only ones, and this was eventually confirmed in the 1950s and 1960s by several writers, including Kurt Heegner (whose proof was incomplete), Harold Stark (who provided a complete proof) and Alan Baker (who proved it independently). We conclude this chapter with their remarkable result:

> *The Baker–Heegner–Stark theorem*: For numbers of the form $a + b\sqrt{n}$, where n is negative and square-free, factorization into primes is unique if and only if
>
> $$n = -1, -2, -3, -7, -11, -19, -43, -67, \text{or} -163.$$

Chapter 8
How to win a million dollars

In 2000 the Clay Mathematics Institute offered a prize of one million US dollars for the solution of each of seven famous problems, widely considered to be among the most important in the subject. The Riemann hypothesis was one of these 'millennium problems', and experts have been trying to prove it for more than 150 years.

So what is the Riemann hypothesis, and why is it important? The problem is concerned with the distribution of prime numbers, and was introduced by Bernhard Riemann, a German mathematician who died at the early age of 39 while a professor at Göttingen University, where he had followed in the footsteps of Gauss and Dirichlet. Elected to the Berlin Academy in 1859, Riemann expressed his gratitude by presenting his only paper in number theory, 'On the number of primes less than a given magnitude' (see Figure 33). Only nine pages long, it is now regarded as a classic.

In very broad terms the Riemann hypothesis asks whether all the solutions of a particular equation have a particular form. This is very vague, and the detailed assertion, which is still unproved, is:

> *Riemann hypothesis*: All the non-trivial zeros of the Riemann zeta function have real part 1/2.

33. (a) Bernhard Riemann.

VII.

Ueber die Anzahl der Primzahlen unter einer gegebenen Grösse.

(Monatsberichte der Berliner Akademie, November 1859.)

Meinen Dank für die Auszeichnung, welche mir die Akademie durch die Aufnahme unter ihre Correspondenten hat zu Theil werden lassen, glaube ich am besten dadurch zu erkennen zu geben, dass ich von der hierdurch erhaltenen Erlaubniss baldigst Gebrauch mache durch Mittheilung einer Untersuchung über die Häufigkeit der Primzahlen; ein Gegenstand, welcher durch das Interesse, welches Gauss und Dirichlet demselben längere Zeit geschenkt haben, einer solchen Mittheilung vielleicht nicht ganz unwerth erscheint.

Bei dieser Untersuchung diente mir als Ausgangspunkt die von Euler gemachte Bemerkung, dass das Product

$$\prod \frac{1}{1-\frac{1}{p^s}} = \Sigma \frac{1}{n^s},$$

wenn für p alle Primzahlen, für n alle ganzen Zahlen gesetzt werden. Die Function der complexen Veränderlichen s, welche durch diese beiden Ausdrücke, so lange sie convergiren, dargestellt wird, bezeichne ich durch $\zeta(s)$. Beide convergiren nur, so lange der reelle Theil von

33. (b) Riemann's 1859 paper.

But what does this mean, and what is its connection with prime numbers?

Infinite series

To investigate these questions we'll need to enter the world of infinite series. The series

$$1 + 1/2 + 1/4 + 1/8 + 1/16 + \ldots,$$

where the denominators are the powers of 2, continues for ever. What happens when we add all these numbers together? Adding them one number at a time gives

$1, \ 1 + 1/2 \, (= 3/2), \ 1 + 1/2 + 1/4 \, (= 7/4),$

$1 + 1/2 + 1/4 + 1/8 \, (= 15/8),$ and so on.

Although we never reach 2 by adding a finite number of terms of the series (see Figure 34), we can get as close to 2 as we wish by adding enough of them—say, the first hundred or the first million: for example, we can get within 0.001 of 2 by adding the first twelve terms. We express this by saying that the infinite series *converges* to 2, or *has the finite sum* 2, and we write

$$1 + 1/2 + 1/4 + 1/8 + 1/16 + \ldots = 2.$$

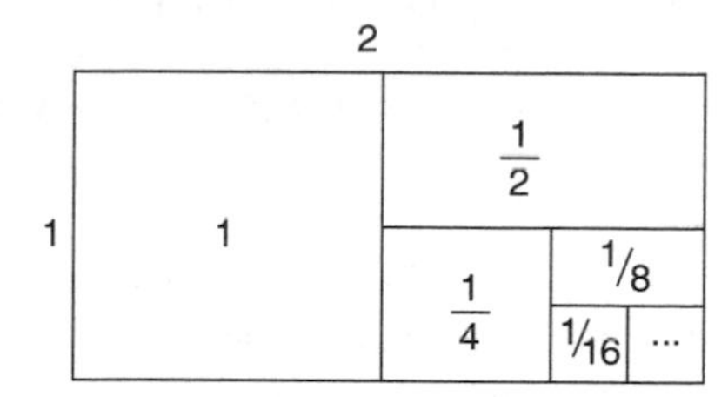

34. Summing the powers of 1/2.

In the same way, we can show that the infinite series

$$1 + 1/3 + 1/9 + 1/27 + \ldots,$$

whose denominators are the powers of 3, converges to 3/2, and that the infinite series

$$1 + 1/5 + 1/25 + 1/125 + \ldots,$$

whose denominators are the powers of 5, converges to 5/4, More generally, we can show that, for any number $p \, (> 1)$

$$1 + 1/p + 1/p^2 + 1/p^3 + \ldots = p/(p-1).$$

We'll need this result later on.

Not all infinite series converge. A celebrated example is the so-called *harmonic series*

$$1 + 1/2 + 1/3 + 1/4 + 1/5 + 1/6 + \ldots,$$

whose denominators are the positive integers. To see why it doesn't have a finite sum we'll group the terms as

$$1 + 1/2 + (1/3 + 1/4) + (1/5 + 1/6 + 1/7 + 1/8) + (1/9 + 1/10 + 1/11 + 1/12 + 1/13 + 1/14 + 1/15 + 1/16) + \ldots.$$

Now this sum is larger than the following sum:

$$1 + 1/2 + (1/4 + 1/4) + (1/8 + 1/8 + 1/8 + 1/8) + (1/16 + 1/16 + 1/16 + 1/16 + 1/16 + 1/16 + 1/16 + 1/16) + \ldots,$$

which equals $1 + 1/2 + 1/2 + 1/2 + 1/2 + \ldots$, because each group of terms has sum 1/2.

But the sum of this latter series increases without limit as we add more terms, and the harmonic series has an even larger sum and so it cannot have a finite sum either.

So the harmonic series doesn't converge: and surprisingly, as Euler proved in 1737, even if we throw away most of its terms and leave only those whose denominators are primes—that is,

$$1 + 1/2 + 1/3 + 1/5 + 1/7 + 1/11 + 1/13 + \ldots$$

—then there is still no finite sum.

The zeta function

In the early 18th century a celebrated challenge was to find the exact sum of the infinite series

$$1 + 1/4 + 1/9 + 1/16 + 1/25 + \ldots,$$

whose denominators are the squares, 1, 4, 9, 16, 25, ... The Swiss mathematician Johann Bernoulli, then possibly the world's greatest mathematician, failed to solve the problem, and it was eventually answered by his former pupil, Leonhard Euler, who proved that this series converges to $\pi^2/6$, a remarkable result in that it involves the 'circle number' π. As Euler proudly observed:

> Quite unexpectedly I have found an elegant formula involving the quadrature of the circle.

In the same way, Euler proved that:

> when the denominators are the fourth powers, the sum is $\pi^4/90$,
>
> when the denominators are the sixth powers, the sum is $\pi^6/945$,
>
> and when the denominators are the eighth powers, the sum is $\pi^8/9450$.

He subsequently continued his calculations up to the twenty-sixth powers. Here the sum is

$$1{,}315{,}862\ \pi^{26}\ /\ 11{,}094{,}481{,}976{,}030{,}578{,}125,$$

which he calculated correctly.

When the denominators are the nth powers, Euler denoted the sum as $\zeta(n)$, where ζ is the Greek letter zeta, and named it the *zeta function*—that is,

$$\zeta(n) = 1 + 1/2^n + 1/3^n + 1/4^n + \ldots.$$

So $\zeta(1)$ is undefined (because the harmonic series has no finite sum), but $\zeta(2) = \pi^2/6$, $\zeta(4) = \pi^4/90$, $\zeta(6) = \pi^6/945$, etc. It turns out that the series for $\zeta(n)$ converges for every number n that is greater than 1.

Although the zeta function $\zeta(n)$ may seem to have nothing in common with prime numbers, Euler spotted a crucial connection, which we now explore. This connection can be used to give another proof that the list of primes is never-ending.

The zeta function and prime numbers

We can write the series for $\zeta(1)$ as follows:

$$\zeta(1) = 1 + 1/2 + 1/3 + 1/4 + 1/5 + 1/6 + \ldots$$

$$= (1 + 1/2 + 1/4 + \ldots) \times (1 + 1/3 + 1/9 + \ldots)$$

$$\times (1 + 1/5 + 1/25 + \ldots) \times \ldots,$$

where each bracket involves the powers of just one prime number. This is because, by unique factorization, each term $1/n$ in the series for $\zeta(1)$ appears exactly once in the product on the right: for example,

$$1/90 = 1/2 \times 1/9 \times 1/5 \times 1 \times 1 \times \ldots$$

$$1/100 = 1/4 \times 1 \times 1/25 \times 1 \times 1 \times \ldots.$$

We now sum the series in each bracket, using our earlier result that

$$1 + 1/p + 1/p^2 + \cdots = p/(p-1):$$

this gives

$$\zeta(1) = 2\,/\,(2-1) \times 3\,/\,(3-1) \times 5/\,(5-1) \times \ldots$$

$$= 2 \times 3/2 \times 5/4 \times \ldots.$$

We can use this product to prove that there are infinitely many primes. For, if there were only finitely many primes, then the right-hand side would be a finite product, and so would have a fixed value. But this would require $\zeta(1)$ to have this same value, which is impossible because it is undefined. So there must be infinitely many primes.

Euler extended these ideas to prove that, for any number n that's greater than 1,

$$\zeta(n) = 2^n\,/\,(2^n - 1) \times 3^n\,/\,(3^n - 1) \times 5^n\,/\,(5^n - 1)$$

$$\times 7^n\,/\,(7^n - 1) \times \ldots;$$

This remarkable result is called the *Euler product*, and provides an unexpected link between the zeta function, which involves powers of numbers and seems to have nothing to do with primes, and a product that intimately involves all the prime numbers. It was a major breakthrough.

Complex numbers

Before we present the Riemann hypothesis, we'll also need the idea of a complex number. This involves i, the 'imaginary' square root of -1, which we met briefly in Chapter 7.
A *complex number* is a symbol of the form $x + yi$; x is called the *real part* of the complex number, and y is the *imaginary part.* Examples of complex numbers are $4 + 3i$, $1/5 - \pi i$, $2i$ (which equals $0 + 2i$), and 3 (which can be thought of as $3 + 0i$).

We can represent complex numbers geometrically as points on the 'complex plane'. This two-dimensional picture consists of all points (x, y), where (x, y) represents the complex number $x + iy$; for example, the points $(4, 3)$, $(1/5, -\pi)$, $(0, 2)$, and $(3, 0)$ represent the complex numbers $4 + 3i$, $1/5 - \pi i$, $2i$, and 3 (see Figure 35).

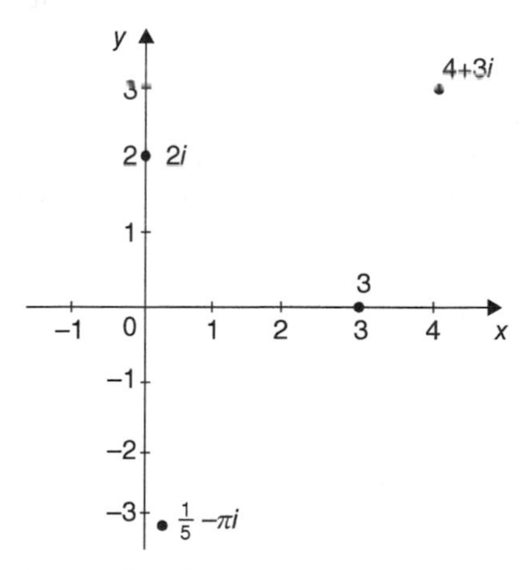

35. Points on the complex plane.

The Riemann hypothesis

We've now set the scene for the Riemann hypothesis.

As we've seen, the zeta function $\zeta(n)$ is defined for any number n that is greater than 1. But can we define it for other numbers n too? For example, how might we define $\zeta(0)$ or $\zeta(-1)$? We can't define them by the same infinite series, because we'd then have

$$\zeta(0) = 1/1^0 + 1/2^0 + 1/3^0 + 1/4^0 + \cdots = 1 + 1 + 1 + 1 + \ldots$$

and

$$\zeta(-1) = 1 + 1/2^{-1} + 1/3^{-1} + 1/4^{-1} + \cdots = 1 + 2 + 3 + 4 + \ldots,$$

and neither of these series has a finite sum. So we'll need to find some other way.

As a clue to how to proceed, we can show that, for certain values of x,

$$1 + x + x^2 + x^3 + \cdots = 1/(1 - x);$$

we've already seen that this is true when $x = 1/p$. But it can be shown that the series on the left-hand side converges only when x lies between -1 and 1, whereas the formula on the right-hand side has a value for any x, apart from 1 (when we get 1/0, which is undefined). So we can *extend* the definition of the series on the left-hand side to *all* values of x (other than 1) by redefining it as the formula on the right-hand side.

In the same way, Riemann found a way of extending the above infinite series definition of the zeta-function to *all* numbers x other than 1 (including 0 and -1). But he went much further than this. Using a technique called 'analytic continuation', Riemann extended the definition of the zeta function to all complex any numbers k other than 1 (because $\zeta(1)$ is undefined) in such a way that when k is a real number greater than 1, we get the same value as before. Because of this, the function is now known as the *Riemann zeta function.*

In Chapter 7 we saw Gauss's attempt to explain why the primes thin out on average, by proposing the estimate $x/\log x$ for the number of primes up to x. Riemann's great achievement was to obtain an *exact formula* for the number of primes up to x, and his formula involved in a crucial way the so-called *zeros of the zeta function*—that is, the complex numbers z that satisfy the equation $\zeta(z) = 0$. But where are these zeros?

It turns out that $\zeta(z) = 0$ when $z = -2, -4, -6, -8, \ldots$; these are called the *trivial zeros* of the zeta function. All the other zeros of the zeta function, the *non-trivial zeros*, are known to lie within a vertical strip between $x = 0$ and $x = 1$ (the so-called *critical strip*), as shown in Figure 36. As we move away from the horizontal axis, the first few non-trivial zeros occur at the following points:

$$1/2 \pm 14.1i, \quad 1/2 \pm 21.01i, \quad \text{and} \quad 1/2 \pm 25.01i.$$

Here, the imaginary parts (such as 14.1) are approximate, but the real parts are all equal to 1/2. Because all of these points all have the form $1/2 \pm$ a multiple of i, the question arises:

Does every zero of the Riemann zeta function in the critical strip lie on the line $x = 1/2$?

The *Riemann hypothesis* is that the answer to this question is 'yes'. It has been proved that the zeros in the critical strip are symmetrically placed, both above and below the x-axis and on either side of the line $x = 1/2$, and that as we progress vertically up and down the line $x = 1/2$, many zeros do lie on it—in fact, *the first trillion zeros lie on this line*! But do *all* of the non-trivial zeros lie on the line $x = 1/2$, or might the first trillion be just a coincidence?

It's now generally believed that all the non-trivial zeros *do* lie on this line, but proving this is one of the most difficult unsolved

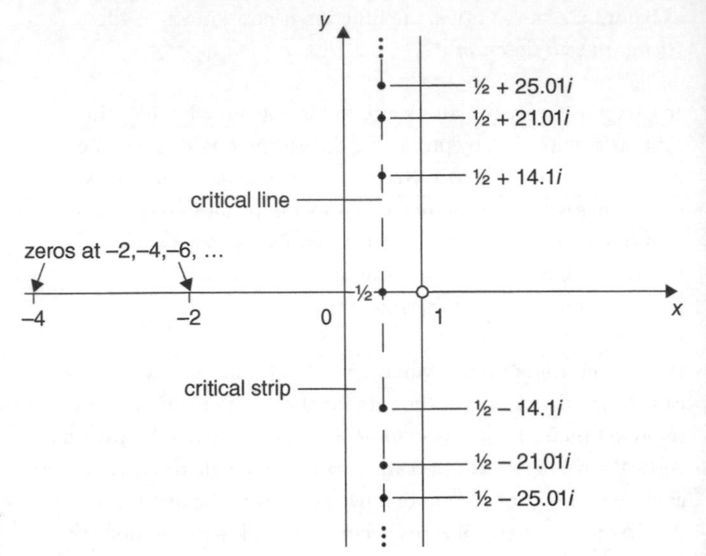

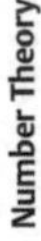

36. The zeros of the Riemann zeta function in the complex plane.

challenges throughout the whole of mathematics. Indeed no-one has yet been able to prove the Riemann hypothesis, even after a century and a half. The million dollar prize is still up for grabs!

Consequences

If the actual statement of the Riemann hypothesis seems an anticlimax after the big build-up, its consequences are substantial. Recalling Riemann's discovery of the role that the zeta function's zeros play in the prime-counting function $\pi(x)$ and in his exact formula (involving the zeta function) for the number of primes up to x, we note that any divergence of these zeros from the line $x = 1/2$ would crucially affect Riemann's exact formula, because our understanding about how the prime numbers behave is so tied up in this formula. Indeed, finding just one zero off the line would cause major havoc in number theory—and in fact throughout mathematics: for a mathematician, truth must be absolute, and admitting even a single exception is forbidden. The prime number theorem would still be true, but would lose its influence on the primes. Instead of Don Zagier's 'military precision', mentioned in Chapter 7, the primes would be found to be in full mutiny!

We conclude this chapter with an unexpected development. In 1972 the American number theorist Hugh Montgomery was visiting the tea-room at Princeton's Institute for Advanced Study and found himself sitting opposite the celebrated physicist Freeman Dyson. Montgomery had been exploring the spacings between the zeros on the critical line, and Dyson said 'But those are just the spacings between the energy levels of a quantum chaotic system'. If this analogy indeed holds, as many think possible, then the Riemann hypothesis may well have consequences in quantum physics. Conversely, using their knowledge of these energy levels, quantum physicists rather than mathematicians may be the ones to prove the Riemann hypothesis. This is truly an intriguing thought!

Chapter 9
Aftermath

In the previous seven chapters we have explored a range of topics from number theory, and we can now return to the questions that we raised in Chapter 1. Most of these questions we were able to answer, while a few others turned out to be famous problems that remain unresolved. For each question we refer back to the chapter in which it was discussed.

The first ten questions

We opened Chapter 1 by posing ten questions.

In which years does February have five Sundays?
Questions that ask for the day of the week on which a particular event falls have a long history, and their solutions often involve the idea of congruence. As we saw in Chapter 4, February has five Sundays in the years 2004, 2032, 2060, and 2088. Also, because 2000 was a leap year, we can find the corresponding years in the previous century by subtracting multiples of 28 years from 2004; those years were 1976, 1948, and 1920. The corresponding years for other centuries can be found in a similar way.

What is special about the number 4,294,967,297*?*
This number is $2^{32} + 1$. As we saw in Chapter 3, Pierre de Fermat believed that all numbers of the form $2^n + 1$, where n is a power of

2, are prime (the so-called 'Fermat primes'), and this number, where $n = 32 = 2^5$, was the smallest that he was unable to check. Years later Leonhard Euler showed that it is not prime, and that 641 is a factor.

How many right-angled triangles with whole-number sides have a side of length 29*?*
Triples of numbers (a, b, c) for which a, b, and c are whole numbers and $a^2 + b^2 = c^2$ are called *Pythagorean triples*, and correspond to side-lengths of right-angled triangles. As we discovered in Chapter 5, there are only two such Pythagorean triples in which one side has length 29: these are (20, 21, 29) and (29, 420, 421).

Are any of the numbers 11, 111, 1111, 11111, ... *perfect squares?*
In Chapter 2 we showed that all perfect squares must have the form $4n$ or $4n + 1$, for some integer n. The numbers in this question all have the form $4n + 3$, and so none of them can be a perfect square.

I have some eggs. When arranged in rows of 3 *there are* 2 *left over, in rows of* 5 *there are* 3 *left over, and in rows of* 7 *there are* 2 *left over. How many eggs are there altogether?*
In Chapter 4 we discussed the solution of simultaneous linear congruences, and gave this problem as an example. It is a version of Sunzi's problem, and asks for the solution of the simultaneous congruences $x \equiv 2 \pmod 3$, $x \equiv 3 \pmod 5$, and $x \equiv 2 \pmod 7$. The smallest solution is $x = 23$, and other answers are found by adding multiples of $105 (= 3 \times 5 \times 7)$.

Can one construct a regular polygon with 100 *sides if measuring is forbidden?*
In Chapter 3 we noted a connection between the construction of regular polygons and the Fermat primes mentioned above: this is Gauss's result that a regular polygon with n sides can be constructed by an unmarked ruler and compasses if and only if n is the product of a power of 2 and unequal Fermat primes. But

$100 = 2^2 \times 5 \times 5$, where the Fermat prime 5 is repeated, and so a regular polygon with 100 sides cannot be so constructed.

How many shuffles are needed to restore the order of the cards in a pack with two Jokers?
Questions involving the shuffling of cards can be answered with help from Fermat's little theorem on primes, as we described in Chapter 6. In this case, where there are 54 cards, the answer is 20 shuffles.

If I can buy partridges for 3 cents, pigeons for 2 cents, and two sparrows for a cent, and if I spend 30 pence on buying 30 birds, how many birds of each kind must I buy?
This is a Diophantine problem requiring an integer solution. As we discussed in Chapter 5, the solution involves two linear equations in three unknowns, together with the extra requirement that the number of each kind of bird that I buy must be a positive integer. The only answer is 5 partridges, 3 pigeons, and 22 sparrows.

How do prime numbers help to keep our credit cards secure?
In Chapter 6 we saw that multiplying large primes is usually a simple matter, whereas trying to factorize large numbers into their prime factors is not. The RSA method of encryption and decryption recognizes this difficulty and involves Euler's theorem, a generalization of Fermat's little theorem.

What is the Riemann hypothesis, and how can I earn a million dollars?
The Riemann hypothesis (or conjecture) was discussed in Chapter 8, and is one of the most famous unsolved problems in mathematics. The conjecture is closely related to that of locating the places where a certain function (called the 'zeta function') has a zero value. It turns out that its proof would tell us much about how prime numbers are distributed, as well as providing its solver with lasting fame and an award of one million dollars by the Clay Mathematics Institute.

Integers

How can we recognize whether a given number, such as 12,345,678, *is a multiple of* 8? *or of* 9? *or of* 11? *or of* 88?

In Chapter 2 we discussed several tests for divisibility by various numbers, such as 8, 9, and 11. The given number is not a multiple of 8 because the number given by its last three digits (678) is not divisible by 8. The sum of its digits is 36, which is divisible by 9, and so the given number is divisible by 9. The alternating sum of its digits is -4, which is not divisible by 11, and so the given number is not divisible by 11 or by any of its multiples, such as 88.

Squares and cubes

We discussed various topics related to squares and higher powers in Chapters 2 and 5. The next three questions were answered in the section on *Squares* in Chapter 2.

Do any squares end in 2, 3, 7, *or* 8?

We showed that all squares must end in 0, 1, 4, 5, 6, or 9, and so there can be no squares that end in 2, 3, 7, or 8.

Must all squares be of the form $4n$ *or* $4n + 1$, *where* n *is an integer?*

As mentioned above, we showed that all squares must be of these forms—in particular, squares of even numbers all have the form $4n$, and squares of odd numbers all have the form $4n + 1$. We can also express this by saying that all squares must be congruent to 0 or 1 (mod 4).

Must the sum of the first few odd numbers, 1, 3, 5, 7, ..., *always be a square?*

For any number k, the first k odd numbers are $1, 3, 5, \ldots, 2k - 1$, and on adding these numbers together we find that their sum is k^2. So the sum of the first few odd numbers is always a square.

The next two questions were discussed in Chapter 5, where we investigated which numbers can be written as the sum of two or more squares.

Which numbers can be written as the sum of two squares?
Because every square is congruent to 0 or 1 (mod 4), the sum of two squares must be congruent to 0, 1, or 2 (mod 4). So no number that is congruent to 3 (mod 4) can be a perfect square. More generally, as stated by Fermat and proved by Legendre, a number can be written as the sum of two squares if and only if every prime factor that is congruent to 3 (mod 4) occurs to an even power.

Can 9999 be written as the sum of two squares? or of three squares? or of four squares?
It follows from our discussions that 9999 cannot be written as the sum of two squares because $9999 \equiv 3 \pmod 4$, or as the sum of three squares because $9999 \equiv 7 \pmod 8$. But, as proved by Lagrange, every positive integer can be written as the sum of four squares: for example, 9999 can be written as $99^2 + 14^2 + 1^2 + 1^2$ or as $90^2 + 43^2 + 7^2 + 1^2$.

Chapter 5 also contained a discussion of which right-angled triangles have sides whose lengths are all integers.

Which right-angled triangles have integer-length sides?
For a right-angled triangle to have integer-length sides, these lengths must be of the form $k\,(x^2 - y^2, 2xy, x^2 + y^2)$, where k is a constant, x and y are coprime integers with one odd and the other even, and $x > y$. Chapter 5 includes a table of all triples with $k = 1$ (the so-called 'primitive triples') and no numbers exceeding 100.

In Chapters 2 and 5 our discussions then extended to cubes.

Must all cubes be of the form $9n$, $9n + 1$, or $9n + 8$, where n is an integer?

The answer to this question is 'yes', as we proved in Chapter 2, at the end of the section on *Squares*. We can also express this result by saying that every cube is congruent to 0, 1, or 8 (mod 9).

Are there any integers a, b, c for which $a^3 + b^3 = c^3$*?*
By Fermat's last theorem, discussed in Chapter 5, the equation $a^n + b^n = c^n$ has no non-zero solutions when $n \geq 3$. So this equation can have solutions only when at least one of the integers a, b, and c is 0: for example, $5^3 + (-5)^3 = 0^3$ and $2^3 + 0^3 = 2^3$.

Can every number be written as the sum of six cubes?
In our discussion of Waring's problem at the end of Chapter 5, we saw that 23, for example, requires at least nine cubes. However, as we remarked, every number from some point onwards can be written as the sum of seven cubes. It is not yet known whether 'seven' can be reduced to 'six'.

Perfect numbers

In Chapter 3 we discussed *perfect numbers*, where a number n is perfect if the sum of all its proper factors (those different from n) is equal to n. The first four perfect numbers, known since the time of the Greeks, are 6, 28, 496, and 8128. The next two questions were discussed in Chapter 3.

What is the next perfect number after 8128?
After 8128 there is a large gap, and the next perfect number does not occur until 33,550,336.

Is there a formula for producing perfect numbers?
As we showed in Chapter 3, every number of the form $2^{n-1} \times (2^n - 1)$, where $2^n - 1$ is prime, is a perfect number, and all *even* perfect numbers can be written in this form—for example,

$$33{,}550{,}336 = 4096 \times 8191 = 2^{12} \times (2^{13} - 1).$$

It is not yet known whether there are any odd perfect numbers.

Prime numbers

Prime numbers were discussed in Chapters 3 and 7. The next two questions were discussed at the beginning of Chapter 7, in the section on *Two famous conjectures.*

Does the list of twin primes go on for ever?
Twin primes are pairs of primes that differ by 2, and many examples are known. The *twin prime conjecture* is that there are infinitely many pairs of twin primes. It is generally believed to be true, but this has never been proved. Some related results are presented in Chapter 7.

Can every even number be written as the sum of two primes?
Another famous unanswered question is *Goldbach's conjecture,* which asks whether every even number greater than 2 can be written as the sum of two primes. It is known to be true for all even numbers up to 400 trillion, but has not yet been proved in general.

In Chapter 7 we also saw how to construct strings of consecutive composite numbers of any desired length.

Is there a string of 1000 *consecutive composite numbers?*
As we saw in the section on *The distribution of primes*, an example of such a string of composite numbers is

$$1001! + 2, \quad 1001! + 3, \quad \ldots, \quad 1001! + 1001.$$

In Chapter 5 we reduced the problem of deciding which numbers can be written as the sum of two squares to the corresponding problem for primes.

Which prime numbers can be written as the sum of two squares?
As we saw in the section on *Sums of squares*, every prime number of the form $4n + 1$ can be written in exactly one way as the sum of two squares, as can the prime 2 ($= 1^2 + 1^2$). However, no numbers

of the form $4n + 3$ (and primes of this form, in particular) can be written as the sum of two squares.

The next two questions were discussed in Chapter 3 in connection with Mersenne and Fermat primes.

Is the number $2^n - 1$ always prime when n is prime, and always composite when n is composite?
In our discussion of Mersenne primes, we saw that $2^n - 1$ must be composite when n is composite. However, $2^n - 1$ need not be prime when n is prime: for example, $2^{11} - 1 = 2047 = 23 \times 89$.

Are all numbers of the form $2^n + 1$, where n is a power of 2, prime?
In our discussion of Fermat primes, we saw that the first five 'Fermat numbers', $2^1 + 1 = 3$, $2^2 + 1 = 5$, $2^4 + 1 = 17$, $2^8 + 1 = 257$, and $2^{16} + 1 = 65,537$, are all prime. No other examples have ever been found.

The last two questions were discussed in Chapter 7, in the section on *Primes in arithmetic progressions.*

Are there infinitely many primes of the form $4n + 1$? or of the form $4n + 3$?
The answer to these questions is 'yes', and both can be proved by adapting Euclid's proof that there are infinitely many primes. In the former case we also used the fact that -1 is a square (mod p) for any prime p of the form $4n + 1$. These results can also be deduced directly from Dirichlet's more general theorem on primes in arithmetic progressions, which states that there are infinitely many prime numbers of the form $an + b$, where n is an integer, as long as $\gcd(a, b) = 1$.

Are there infinitely many primes with final digit 9?
Again, by Dirichlet's theorem, there are infinitely many prime numbers of the form $10n + 9$—that is, primes with final digit 9.

We have now reached the end of our story. Number theory continues to be an exciting part of modern mathematics, with many startling new developments over recent years. However, there are many parts of the subject that we have been unable to explore within these pages, and we hope that you will wish to continue your interest in the subject by referring to the items in our list of further reading.

Further reading

The following texts, some classic and others much newer, provide useful introductions to the various areas of number theory introduced in this book.

George E. Andrews, *Number Theory*, new edn, Dover Publications, 2000.

David M. Burton, *Elementary Number Theory*, 7th edn, McGraw-Hill, 1980.

H. Davenport, *The Higher Arithmetic*, 8th edn, Cambridge University Press, 2008.

Underwood Dudley, *Elementary Number Theory*, 2nd edn, Dover Publications, 2008.

Emil Grosswald, *Topics from the Theory of Numbers*, 2nd edn, Birkhäuser, 1984.

G. H. Hardy and E. M. Wright, *An Introduction to the Theory of Numbers*, 6th edn (ed. D. R. Heath-Brown and J. H. Silverman), Oxford University Press, 2008.

Gareth A. Jones and J. Mary Jones, *Elementary Number Theory*, Springer, 1998.

Oystein Ore, *Invitation to Number Theory*, 2nd edn (revised and updated by John J. Watkins and Robin Wilson), Mathematical Association of America, 2017.

James J. Tattersall, *Elementary Number Theory in Nine Chapters*, 2nd edn, Cambridge University Press, 2005.

Martin H. Weissman, *An Illustrated Theory of Numbers*, American Mathematical Society, 2017.

More historical information can be found in:
Leonard Eugene Dickson, *History of the Theory of Numbers*, Vols. I, II, III, Dover Publications, 2005.
O. Ore, *Number Theory and its History*, Dover Publications, 1988.
John J. Watkins: *Number Theory: A Historical Approach*, Princeton University Press, 2014.

Popular books on the Riemann hypothesis, Fermat's last theorem, and the twin prime conjecture are:
John Derbyshire, *Prime Obsession: Bernhard Riemann and the Greatest Unsolved Problem in Mathematics*, Plume Books, 2004.
Marcus du Sautoy, *The Music of the Primes: Why an Unsolved Problem in Mathematics Matters*, Harper/Perennial, 2004.
Vicky Neale, *Closing the Gap: The Quest to Understand Prime Numbers*, Oxford University Press, 2017.
Karl Sabbagh, *Dr Riemann's Zeros*, Atlantic Books, 2003.
Simon Singh, *Fermat's Last Theorem*, Fourth Estate, 1998.

More advanced books on specific topics include:
Tom M. Apostol, *Introduction to Analytic Number Theory*, 5th printing, Springer, 1998.
H. M. Edwards, *Riemann's Zeta Function*, Dover Publications, 2003.
Barry Mazur and William Stein, *Prime Numbers and the Riemann Hypothesis*, Cambridge University Press, 2016.
Paulo Ribenboim, *The Little Book of Bigger Primes*, 2nd edn, Springer, 2010.
Ian Stewart and David Tall, *Algebraic Number Theory and Fermat's Last Theorem*, 3rd edn, A. K. Peters / CRC Press, 2001.

An English translation of Gauss's classic book of 1801 is:
Carl Friedrich Gauss, *Disquisitiones Arithmeticae* (trans. Arthur A. Clarke), Yale University Press, 1965.

Index

A

B

C

D

E

F

G

H

I

J

K

L

M

N

O

P

Economics

A Very Short Introduction

Partha Dasgupta

Economics has the capacity to offer us deep insights into some of the most formidable problems of life, and offer solutions to them too. Combining a global approach with examples from everyday life, Partha Dasgupta describes the lives of two children who live very different lives in different parts of the world: in the Mid-West USA and in Ethiopia. He compares the obstacles facing them, and the processes that shape their lives, their families, and their futures. He shows how economics uncovers these processes, finds explanations for them, and how it forms policies and solutions.

> 'An excellent introduction . . . presents mathematical and statistical findings in straightforward prose.'
>
> Financial Times

www.oup.com/vsi

INFORMATION

A Very Short Introduction

Luciano Floridi

Luciano Floridi, a philosopher of information, cuts across many subjects, from a brief look at the mathematical roots of information - its definition and measurement in 'bits'- to its role in genetics (we are information), and its social meaning and value. He ends by considering the ethics of information, including issues of ownership, privacy, and accessibility; copyright and open source. For those unfamiliar with its precise meaning and wide applicability as a philosophical concept, 'information' may seem a bland or mundane topic. Those who have studied some science or philosophy or sociology will already be aware of its centrality and richness. But for all readers, whether from the humanities or sciences, Floridi gives a fascinating and inspirational introduction to this most fundamental of ideas.

'Splendidly pellucid.'

Steven Poole, The Guardian

www.oup.com/vsi

INNOVATION
A Very Short Introduction

Mark Dodgson & David Gann

This *Very Short Introduction* looks at what innovation is and why it affects us so profoundly. It examines how it occurs, who stimulates it, how it is pursued, and what its outcomes are, both positive and negative. Innovation is hugely challenging and failure is common, yet it is essential to our social and economic progress. Mark Dodgson and David Gann consider the extent to which our understanding of innovation developed over the past century and how it might be used to interpret the global economy we all face in the future.

> 'Innovation has always been fundamental to leadership, be it in the public or private arena. This insightful book teaches lessons from the successes of the past, and spotlights the challenges and the opportunities for innovation as we move from the industrial age to the knowledge economy.'
>
> Sanford, Senior Vice President, IBM

www.oup.com/vsi

NOTHING

A Very Short Introduction

Frank Close

What is 'nothing'? What remains when you take all the matter away? Can empty space - a void - exist? This *Very Short Introduction* explores the science and history of the elusive void: from Aristotle's theories to black holes and quantum particles, and why the latest discoveries about the vacuum tell us extraordinary things about the cosmos. Frank Close tells the story of how scientists have explored the elusive void, and the rich discoveries that they have made there. He takes the reader on a lively and accessible history through ancient ideas and cultural superstitions to the frontiers of current research.

> 'An accessible and entertaining read for layperson and scientist alike.'
>
> **Physics World**

www.oup.com/vsi